J. G. Dif

Computational Methods for Matrix Eigenproblems

Computational Methods for Matrix Eigenproblems

A. R. GOURLAY

IBM United Kingdom Scientific Centre
Winchester, Hampshire

G. A. WATSON

Department of Mathematics
University of Dundee

JOHN WILEY & SONS

Chichester · New York · Brisbane · Toronto

Library of Congress Catalog Card No. 73-2783

ISBN 0 471 27586 7

Produced by offset lithography by
UNWIN BROTHERS LIMITED
The Gresham Press, Old Woking, Surrey

To
Wendy and Hilary

Preface

Since 1947, numerical analysis has enjoyed dedicated effort from mathematicians (both pure and applied), computer scientists, engineers and the like, all striving to devise and analyse processes aimed at being of use and value in the solution of the problems of the outside world. One of the topic areas of numerical analysis in which this effort has led to rich rewards is that of computational linear algebra. Our understanding of matrix problems is much more complete than that of almost any other subset of numerical analysis. We have powerful algorithms for solving linear equations, and for tackling nearly all the commonly occurring eigenproblems. More important perhaps is the existence of reliable and robust computer routines for these algorithms. Much of this knowledge is contained in the monumental work by Wilkinson (*The Algebraic Eigenvalue Problem*, O.U.P., 1965). The rest is in the more recent literature.

The topics, numerical solution of linear systems, and numerical solution of matrix eigenproblems, are fundamental to courses in numerical analysis. Therefore, the availability of suitable textbooks for such courses is desirable. For the first topic, the book by Forsythe and Moler (*Computer Solution of Linear Systems*, Prentice Hall, 1967) is very suitable. Its style of short, single topic chapters is one which should appeal to lecturers and students alike. The present text is an attempt to provide a similar book for the second topic.

The content is based on lectures given by the authors to M.Sc. students at the University of Dundee. Some of the material has also been used in undergraduate courses at the same University. All these students would have attended a basic course in matrix algebra and be familiar with the fundamental concepts, for example the definitions of matrix multiplication, inversion and determinants. This book is written for students with such a background. The material is suitable for courses to students both in mathematical disciplines and in the more applied subjects such as engineering, and, because of the particular format in which the material is presented, it could form the basis for any level of course in its subject matter. Each chapter could be covered in one or two lectures, and thus the whole text could be suitable for a course of some fifteen to twenty lectures (including an allocation of time for consideration of some of the relevant chapters in the book by Forsythe and Moler).

The subjects have been chosen so as to present only the more commonly used and more reliable techniques for computing solutions to eigenproblems.

The aim is primarily to describe the techniques; therefore little will be said of the error analysis of each method, although the conclusions to be drawn from the relevant error analyses will of course be stressed. It is hoped that any student whose interest may be roused by the material covered here will turn to the book by Wilkinson for the 'complete story'.

Many colleagues and friends have contributed to the development of this book, particularly in the early stages of preparation. One of the authors (ARG) is especially grateful to Dr. J. Ll. Morris for his helpful criticism of first drafts of several of the chapters, to Professor D. S. Jones not only for the initial encouragement to undertake the preparation of this text but also for arranging the rescue of a foundering manuscript, and to his co-author (GAW) for effecting the rescue.

Both authors wish to express their thanks to Professor D. S. Jones for his advice and comments throughout the preparation of the manuscript, and to Mrs. Hilary Watson and Miss Frances Duncan for the expert typing of the manuscript.

A. R. GOURLAY
G. A. WATSON

Contents

1

Introduction

1.1 Introduction

The aim of this book is to provide an elementary text on the numerical techniques used in the solution of the algebraic eigenproblem

$$A\mathbf{x} = \lambda\mathbf{x}$$

where A is a known $n \times n$ matrix. The scalar λ is referred to variously as eigenvalue, latent root, characteristic value of the matrix A and the n column vector $\mathbf{x}$ as an eigenvector, latent vector, characteristic vector of A. The purpose of this introductory chapter is to provide some examples of the occurrence of such problems in a wide variety of areas of application. Certain standard definitions and conventions of notation are assumed, and the reader unfamiliar with these should refer to Chapter 2, which contains a collection of the standard notations and more useful results (to this text). Each of the following sections is intended to be self-contained and independent.

1.2 A geometrical example

The equation of an ellipsoid in n space dimensions is given, in cartesian form, by

$$\tfrac{1}{2}\sum_{i=1}^{n}\sum_{j=1}^{n} a_{ij}z_i z_j + \sum_{j=1}^{n} b_j z_j + c' = 0,$$

where $\mathbf{z} = (z_1, z_2, \ldots, z_n)^T$ is a point in n dimensional space. Using matrix notation this may be compactly written as

$$\tfrac{1}{2}\mathbf{z}^T A\mathbf{z} + \mathbf{b}^T\mathbf{z} + c' = 0,$$

where c' is a known constant, $\mathbf{b}$ is a known vector and A a known symmetric positive definite matrix. By a suitable translation of axes

$$\mathbf{z} = \mathbf{x} - A^{-1}\mathbf{b},$$

the equation may be simplified to

$$\tfrac{1}{2}\mathbf{x}^T A\mathbf{x} + c = 0.$$

If the position vector of a point $\mathbf{x}$ on this *hyperellipsoid* is the same as the gradient vector at $\mathbf{x}$, then $\mathbf{x}$ is a principal axis of the hyperellipsoid. Thus the set of principal axes are those directions which simultaneously correspond to a position vector and to a gradient vector. A two-dimensional example will help to clarify this. (Here the two components of $\mathbf{x}$ are denoted by x and y.)

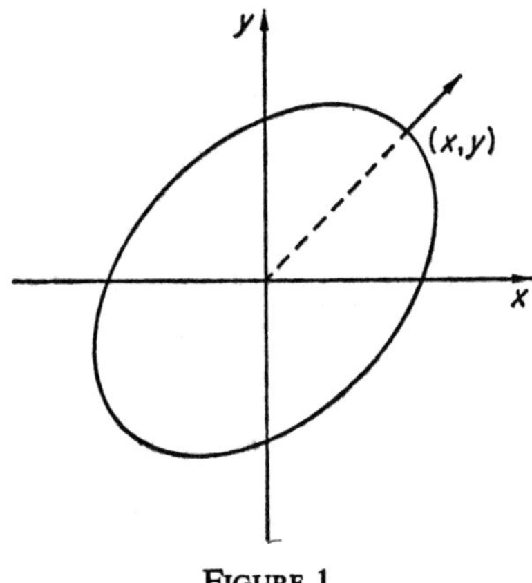

FIGURE 1

In Figure 1 the curve

$$ax^2+2bxy+cy^2 = d$$

is drawn. At a general point (x, y), the gradient is in the direction

$$(ax+by, bx+cy).$$

At the particular point (x, y) in Figure 1 the gradient vector is in the same direction as the position vector from the origin. It follows that at this point (x, y) (and at any similar points) there exists some scalar λ, such that

$$ax+by = \lambda x$$

$$bx+cy = \lambda y,$$

or in matrix notation

$$\begin{bmatrix} a & b \\ b & c \end{bmatrix} \begin{bmatrix} x \\ y \end{bmatrix} = \lambda \begin{bmatrix} x \\ y \end{bmatrix}.$$

From this equation we deduce that the principal axes are given by the eigenvectors of the matrix

$$\begin{bmatrix} a & b \\ b & c \end{bmatrix}.$$

Returning to our n dimensional example we observe that the gradient vector is given by

$$\mathbf{g} = A\mathbf{x}.$$

Thus the principal axes are given by the n (non-trivial) vectors $\mathbf{x}$ satisfying

$$A\mathbf{x} = \lambda\mathbf{x},$$

that is by the n eigenvectors of A.

1.3 Small vibrations

An area which is a fruitful source of eigenproblems is the study of the vibrations of dynamical, and structural systems. The example given below considers the small vibrations of particles on a string under tension. Simplifying assumptions have been made to ensure that the analysis does not become too complicated. Thus we assume a uniform weightless string, no gravity and that the vibrations are small and in a direction perpendicular to the rest position of the string. We consider specifically the motion of four unequal, but equally spaced, particles on a string under tension F. The system is shown in Figure 2.

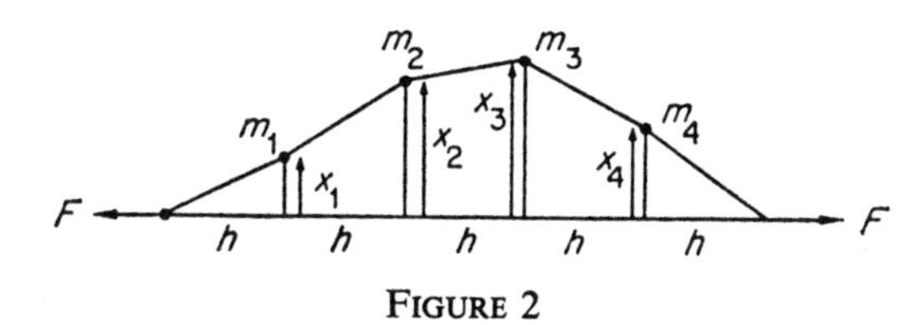

FIGURE 2

Making the standard assumptions, the equations for this system are given by

$$m_1 \frac{d^2x_1}{dt^2} = -F\frac{x_1}{h} + F\left(\frac{x_2 - x_1}{h}\right)$$

$$m_2 \frac{d^2x_2}{dt^2} = -F\left(\frac{x_2 - x_1}{h}\right) + F\left(\frac{x_3 - x_2}{h}\right)$$

$$m_3 \frac{d^2x_3}{dt^2} = -F\left(\frac{x_3 - x_2}{h}\right) - F\left(\frac{x_3 - x_4}{h}\right)$$

$$m_4 \frac{d^2x_4}{dt^2} = +F\left(\frac{x_3 - x_4}{h}\right) - F\frac{x_4}{h}.$$

Defining the vector $\mathbf{x} = (x_1, x_2, x_3, x_4)^T$ and letting

$$d_i = \frac{m_i h}{F}, \quad i = 1, 2, 3, 4,$$

this system may be written in matrix notation as

$$D\frac{d^2\mathbf{x}}{dt^2} = T\mathbf{x} \tag{1}$$

where D is the diagonal matrix

$$D = \begin{bmatrix} d_1 & 0 & 0 & 0 \\ 0 & d_2 & 0 & 0 \\ 0 & 0 & d_3 & 0 \\ 0 & 0 & 0 & d_4 \end{bmatrix}$$

and T is the tridiagonal matrix

$$T = \begin{bmatrix} -2 & 1 & 0 & 0 \\ 1 & -2 & 1 & 0 \\ 0 & 1 & -2 & 1 \\ 0 & 0 & 1 & -2 \end{bmatrix}.$$

When the system vibrates in a normal mode the equation

$$\frac{d^2\mathbf{x}}{dt^2} = -w^2\mathbf{x} \tag{2}$$

holds. (In this situation the masses all vibrate in phase or in direct opposition.) Substituting (2) in (1), we obtain the eigenproblem

$$Dw^2\mathbf{x} = -T\mathbf{x} \tag{3}$$

for the normal frequencies $w_1, \ldots, w_4$ and the corresponding normal modes. Although this would appear at first sight to be a *generalized eigenproblem* of the form

$$(A - \lambda B)\,\mathbf{x} = 0,$$

it may easily be transformed into the standard symmetric tridiagonal problem

$$D^{-1/2}TD^{-1/2}\mathbf{y} = -w^2\mathbf{y}$$

where $\mathbf{y} = D^{1/2}\mathbf{x}$, since the elements of D are positive.

This model may easily be extended to the general case of n particles on a string, leading to an n dimensional analogue of (3). The matrix T is still tridiagonal. In fact, it is a special matrix which occurs frequently in numerical analysis, and whose eigenvalues are expressible in analytic terms.

1.4 An example in information system design

If we regard an information (both storage and retrieval) system as made up of component subsystems which operate together and perform a set of operations to accomplish the defined purpose of the system, then the aims in the design of such a system may be stated as:

(i) to define the purpose of the system
(ii) to select the component subsystems to achieve this purpose in an optimal way.

In the following analysis of this problem, we shall see that the eigenvectors of a particular matrix play an important role. We begin with a few definitions used in the model.

A *job* is defined to be the purpose of the system and it is composed of a set of *operations* $O_1, \ldots, O_m$ together with a *volume* (or work load) $V_1, \ldots, V_m$ for each operation. A *component* is a well-defined means of performing some parts of these operations. The *efficiency* with which a given component performs the execution of a given operation is measured as a function of cost, time and size of the operation. For example, a typical choice would be

$$e = \frac{ct}{n}$$

where c is the cost in pounds per unit time, t is the time taken and n is a measure of the size of the operation (e.g. bits processed). A *total system* constructed from a set of components $(S_1, \ldots, S_n)$ in order to achieve all the required operations is represented by an $n \times m$ efficiency matrix E with the (i, j) element e_{ij}, where for example,

$$e_{ij} = \frac{c_{ij} t_{ij}}{n_{ij}},$$

is the efficiency with which the ith component performs operation O_j.

Since the above system is designed to perform a specified job made up of the operations together with a volume for each operation, the cost of this system performing the job is

$$\mathbf{x} = E\mathbf{v},$$

where

$$\mathbf{x} = (x_1, x_2, \ldots, x_n)^T$$

$$\mathbf{v} = (v_1, v_2, \ldots, v_m)^T.$$

The system and its performance are thus given by E, $\mathbf{v}$ and $\mathbf{x}$. We require some measure of the performance of the system on various tasks, defined by different volume vectors $\mathbf{v}$. One measure is the *Rayleigh quotient*

$$a^2 = \frac{(E\mathbf{v})^T (E\mathbf{v})}{\mathbf{v}^T\mathbf{v}}.$$

The design of a general or total system involves the calculation of the maximum cost of the system, and the maximum of the above measure is given by the largest eigenvalue of E^TE. Further, the corresponding value of $\mathbf{v}$ giving the critical volume at which this maximum is achieved is given by the corresponding eigenvector of E^TE.

Further details of this area of application can be found, for example, in the book by Becker and Hayes (1967).

1.5 An eigenproblem in non-linear optimization

A basic problem in non-linear optimization is the determination of the n dimensional vector $\mathbf{x}$ which minimizes the scalar function $f(\mathbf{x}) = f(x_1, \ldots, x_n)$. Assuming that we are able to calculate the gradient of $f(\mathbf{x})$, denoted by $\mathbf{g}(\mathbf{x})$, then we may use a member of the class of *variable metric methods*. These algorithms assume an initial guess or estimate $\mathbf{x}_0$ of the solution and calculate a sequence of new points $\{\mathbf{x}_k\}$ by means of a relation of the form

$$\mathbf{x}_{k+1} = \mathbf{x}_k + t_k\mathbf{d}_k$$

where $\mathbf{d}_k$ is a direction vector and t_k is a positive scalar chosen to minimize

$$f(\mathbf{x}_k + t\mathbf{d}_k)$$

with respect to t—a univariate minimization problem. The variable metric methods are characterized by the use of a direction vector of the form

$$\mathbf{d}_k = -H_k\mathbf{g}_k$$

where $\mathbf{g}_k = \mathbf{g}(\mathbf{x}_k)$ and H_k is a symmetric positive definite matrix. It is beyond the scope of this section to explain in detail the theory of such algorithms.

At each step a new approximate matrix H_{k+1} is computed by a relation of the form

$$H_{k+1} = H_k + E_k, \qquad H_0 = I.$$

In practice, it is essential to ensure that the sequence of matrices $\{H_k\}$ remains positive definite. Whilst the correction E_k is usually chosen so that H_{k+1} will be positive definite if H_k is positive definite, the presence of rounding errors may cause H_{k+1} to become indefinite. J. Greenstadt has suggested a means of ensuring positive definiteness which involves a complete eigenanalysis of

H_{k+1}. If $\{\lambda_i^{k+1}\}$ and $\{\mathbf{u}_i^{k+1}\}$ are the eigenvalues and orthonormal eigenvectors of H_{k+1} then, we may write

$$H_{k+1} = \sum_{i=1}^{n} \lambda_i^{k+1}\mathbf{u}_i^{k+1}\{\mathbf{u}_i^{k+1}\}^T.$$

We now redefine H_{k+1} to be

$$H_{k+1} = \sum_{i=1}^{n} |\lambda_i^{k+1}|\, \mathbf{u}_i^{k+1}\{\mathbf{u}_i^{k+1}\}^T$$

which ensures that H_{k+1} is non-negative definite. If however any member of the set $\{\lambda_i^{k+1}\}$ were zero then the safest strategy would be to define

$$H_{k+1} = I.$$

This suggestion of Greenstadt, whilst ensuring positive definiteness, unfortunately involves a considerable increase in the computational requirements of the algorithms. For this reason it is only feasible for problems with small dimension n.

1.6 An example from mathematical economics

In the study of macroeconomics, one of the most useful tools available to the planner is input–output analysis introduced by Leontief. The input–output table or Leontief matrix links the individual industries to the overall working of the economy. To introduce the concepts we follow the book of Dernburg and Dernburg (1969).

Considering the sales and purchases of an industrial sector, we denote by b_{ij} the sales of industry i to industry j, and by b_{ii} the retention of goods produced by industry i. The sales of goods produced by industry i to outside users is denoted by y_i and the gross output by x_i. Thus

$$x_i = y_i + \sum_j b_{ij}. \tag{4}$$

The next step is to define the input coefficient. We assume that the sales of industry i to industry j are in constant proportion (a_{ij}) to the output of industry j, thus

$$b_{ij} = a_{ij}x_j.$$

The quantities a_{ij} are defined to be the input coefficients. From equation (4) we see that in a static situation

$$\mathbf{x} = \mathbf{y} + A\mathbf{x}, \tag{5}$$

where

$$\mathbf{x} = (x_1, x_2, \ldots, x_n)^T,$$

$$\mathbf{y} = (y_1, y_2, \ldots, y_n)^T,$$

and A is the $n \times n$ matrix with the (i, j) element a_{ij}. The matrix $(I-A)$ is known as the Leontief matrix. Equation (5) can be used to determine the required gross outputs $\mathbf{x}$ of the industry sector to meet a preset final demand $\mathbf{y}$.

If supply and demand are not in equilibrium then we must replace equation (5) by a dynamic model. The usual assumption is that the output in each industry changes at a rate which is proportional to the difference between the level of sales and the level of production. Thus our dynamic model takes the form

$$\frac{\mathrm{d}\mathbf{x}(t)}{\mathrm{d}t} = D[(A-I)\,\mathbf{x}(t)+\mathbf{y}(t)], \tag{6}$$

where D is a diagonal matrix of the reaction coefficients of the industries. Equation (6) thus is a simple model of the dynamic behaviour of the economic system we are considering. The question of the stability of the system being modelled can now be answered, by determining the eigensystem of the matrix $D(A-I)$ and thus considering the behaviour of the solutions to the system (6). In particular, for this model the existence of eigenvalues with positive real part would indicate an instability in the system because the required gross output would grow exponentially with time.

A similar use of the Leontief matrix and eigensystem analysis occurs in a discrete dynamic system of the form

$$\mathbf{x}(t+1)-\mathbf{x}(t) = D[(A-I)\,\mathbf{x}(t)+\mathbf{y}(t)].$$

Such models are of use in studying the stability of interindustry relations, multiple markets and intercountry trade. For fuller details the reader is referred to the text of Dernburg and Dernburg (1969).

1.7 A Sturm–Liouville problem

In the numerical analysis of ordinary and partial differential equations, a commonly occurring problem is the determination of an approximating eigensystem of the continuous problem. This may represent the vibration of bars, plates or structures, the oscillation of fluids, etc. Many of these problems are now tackled by variational means using a technique frequently referred to as a finite element or Rayleigh–Ritz method. Our example in this section is a straightforward Rayleigh–Ritz attack on a Sturm–Liouville problem. Our aim is to demonstrate the technique and the resulting eigenproblem in as simple a manner as possible.

We therefore consider the problem of determining those values of λ for which there exists a non-trivial differentiable function $\phi(x)$ on $[a, b]$ which, under suitable assumptions, satisfies the differential equation

$$(p(x)\phi'(x))'-q(x)\phi(x)+\lambda r(x)\phi(x) = 0, \tag{7}$$

the boundary conditions

$$\phi(a) = \phi(b) = 0, \tag{8}$$

and the normalization condition

$$\int_a^b r(x)\phi^2(x)\,\mathrm{d}x = 1. \tag{9}$$

The functions $p(x)$, $q(x)$, $r(x)$ satisfy $p(x)>0$, $q(x)\geqslant 0$ and $r(x)>0$. The next step in the analysis is to place a mesh on the interval $[a, b]$ consisting of the points

$$a = x_0 < x_1 < x_2 < \ldots < x_{N+1} = b.$$

Let M denote a subspace of functions defined on the mesh on the interval $[a, b]$. For example, M might be chosen to be a *spline* subspace such that each $\psi_j \in M$ is a cubic polynomial on each interval $[x_j, x_{j+1}]$, $j = 0, \ldots, N$ and such that ψ_j has continuous second derivative at the points x_j. Practical considerations regarding the choice of M and the basis functions ψ_j are beyond the scope or intention of this example.

Returning to our Sturm–Liouville problem we cast our eigenproblem in the form of a Rayleigh–Ritz minimization. Thus the solution to (7), (8) and (9) is equivalent to finding the stationary values, and corresponding functions ϕ of the Rayleigh quotient

$$R[\phi] = \left\{\int_a^b [p(\phi')^2 + q\phi^2]\,\mathrm{d}x\right\} \Big/ \left\{\int_a^b r\phi^2\,\mathrm{d}x\right\}. \tag{10}$$

In general, we cannot deal with (10) unless we make some simplifying assumptions. If we restrict our approximate solution ϕ to lie in the (usually finite) subspace M then we may carry the analysis further. Thus letting

$$\phi = \sum_{j=1}^{J} c_j\psi_j(x)$$

where c_j are constants to be determined, the problem reduces to that of determining the values of c_j corresponding to the stationary values of (10). For ease of writing we use the notation

$$\phi = \mathbf{c}^T\boldsymbol{\psi} \tag{11}$$

where

$$\mathbf{c} = (c_1, c_2, \ldots, c_J),$$

$$\boldsymbol{\psi} = (\psi_1, \psi_2, \ldots, \psi_J).$$

If we substitute into (10) the assumption (11), then

$$R[\mathbf{c}] = N(\mathbf{c})/D(\mathbf{c}),$$

where

$$N(\mathbf{c}) = \int_a^b p(x)(\mathbf{c}^T \psi'(x))^2 + q(x)(\mathbf{c}^T \psi(x))^2 \, dx$$

$$D(\mathbf{c}) = \int_a^b r(x)(\mathbf{c}^T \psi(x))^2 \, dx.$$

The stationary values of this approximate Rayleigh quotient are obtained by differentiation with respect to the unknown vector $\mathbf{c}$. The resulting equation obtained is

$$\mathbf{0} = \frac{1}{D(\mathbf{c})^2} [D(\mathbf{c})\nabla_c N(\mathbf{c}) - N(\mathbf{c})\nabla_c D(\mathbf{c})].$$

If $\hat{\mathbf{c}}$ is a solution to this system then it is given by

$$\nabla_c N(\hat{\mathbf{c}}) - \frac{N(\hat{\mathbf{c}})}{D(\hat{\mathbf{c}})} \nabla_c D(\hat{\mathbf{c}}) = \mathbf{0}$$

or

$$\nabla_c N(\hat{\mathbf{c}}) - R(\hat{\mathbf{c}})\nabla_c D(\hat{\mathbf{c}}) = \mathbf{0}. \tag{12}$$

$R(\hat{\mathbf{c}})$ is the stationary value of $R(\phi)$ for this solution $\hat{\mathbf{c}}$, and is an approximation to an eigenvalue of (7) with corresponding *eigenfunction* given by

$$\phi = \hat{\mathbf{c}}^T \psi.$$

Now since

$$\nabla_c N(\mathbf{c}) = 2 \int_a^b \{p(\mathbf{c}^T \psi') \, \psi' + q(\mathbf{c}^T \psi) \, \psi\} \, dx,$$

$$\nabla_c D(\mathbf{c}) = 2 \int_a^b r(\mathbf{c}^T \psi) \, \psi \, dx$$

are both linear in $\mathbf{c}$, it follows that (12) is a linear matrix eigenproblem of the form

$$A\mathbf{x} - \lambda B\mathbf{x} = 0.$$

Indeed, the matrices A, B in this example are symmetric and positive definite. Moreover for appropriate choice of the subspace M, both A and B may be narrow band matrices (i.e. the non-zero elements are grouped close to the main diagonal). This type of eigenproblem is dealt with in Chapter 14.

Thus the procedure which we have developed has replaced the determination of approximations to a Sturm–Liouville eigensystem by a matrix eigenproblem.

2

Background theory

2.1 Introduction

We now give an outline of the matrix theory which will be required in later chapters. Most of the basic results are proved; however, in the interests of compactness, we have generally avoided giving proofs which introduce material which is not subsequently used.

It is convenient to introduce some conventions of notation which we shall follow. Matrices will be denoted by capital letters, and a matrix A referred to as being $m \times n$ will have m rows and n columns, with the (i, j) element a_{ij}, $i = 1, 2, \ldots, m; j = 1, 2, \ldots, n$. If $m = n$, A is said to be *square*, and in this case we denote its determinant by det (A). If det $(A) = 0$, A is said to be *singular*; otherwise it is *non-singular*.

The identity matrix will be denoted by I. The $n \times n$ matrix with the (i, j) element equal to $\lambda_i \delta_{ij}$, where δ_{ij} is the Kronecker delta satisfying

$$\delta_{ij} = 0,\ i \neq j;\ \delta_{ii} = 1,\ i, j = 1, 2, \ldots, n,$$

is called a *diagonal matrix*, and will be represented by diag (λ_i). If the (i, j) element is zero for all i, j such that $|i-j| > 1$ (where $|a|$ represents the modulus of any number a), then the matrix is said to be *tridiagonal.* The square matrix with the (i, j) element equal to zero for $i > j (i < j)$ is called an *upper* (*lower*) *triangular matrix*, and the square matrix with the (i, j) element equal to zero for $i > j+1$ $(i+1 < j)$ is called an *upper* (*lower*) *Hessenberg matrix.*

Examples of diagonal, tridiagonal, upper triangular and lower Hessenberg matrices are respectively:

$$(a)\quad \begin{bmatrix} 1 & 0 & 0 & 0 \\ 0 & 2 & 0 & 0 \\ 0 & 0 & 3 & 0 \\ 0 & 0 & 0 & 4 \end{bmatrix} \qquad (b)\quad \begin{bmatrix} -1 & 5 & 0 & 0 \\ 2 & 1 & 6 & 0 \\ 0 & 3 & 3 & -1 \\ 0 & 0 & 4 & 1 \end{bmatrix}$$

$$(c)\ \begin{bmatrix} 1 & 2 & 3 & 4 \\ 0 & 5 & 6 & 7 \\ 0 & 0 & 8 & 9 \\ 0 & 0 & 0 & 10 \end{bmatrix} \qquad (d)\ \begin{bmatrix} 2 & 2 & 0 & 0 \\ 4 & 2 & 1 & 0 \\ -1 & -1 & 2 & 3 \\ 1 & 3 & 5 & 2 \end{bmatrix}$$

When a matrix being displayed has a number of zero entries occurring together in a block, it is convenient to omit explicit representation of the individual zeros, and merely to leave blanks. Thus, we usually display the matrices of, for example, (*a*) and (*c*) as

$$\begin{bmatrix} 1 & & & \\ & 2 & & \\ & & 3 & \\ & & & 4 \end{bmatrix} \quad \text{and} \quad \begin{bmatrix} 1 & 2 & 3 & 4 \\ & 5 & 6 & 7 \\ & & 8 & 9 \\ & & & 10 \end{bmatrix}$$

respectively. A matrix with a large number of zero elements is said to be *sparse*. If there are few zero elements, the matrix is said to be *dense*.

The transpose of A, denoted by A^T, is the matrix with the (i, j) element equal to a_{ji}. The matrix with the (i, j) element $\bar{a}_{ij}$, the complex conjugate of a_{ij}, is written $\bar{A}$, and A^* is used to denote the matrix with the (i, j) element $\bar{a}_{ji}$. It follows that $A^* = \bar{A}^T$.

Vectors will be represented by bold face lower case letters, with $\mathbf{x}$ denoting a column vector with elements $x_1, x_2, \ldots, x_n$ for some n. The row vector with the same elements is denoted by $\mathbf{x}^T$, and the row vector with elements $\bar{x}_1, \bar{x}_2, \ldots, \bar{x}_n$ by $\mathbf{x}^*$. We will have occasion to use vectors with 1 in the ith position and zeros elsewhere; these will be denoted by $\mathbf{e}_i$, e.g. $\mathbf{e}_2 = (0, 1, 0, \ldots, 0)^T$. Where vectors (or matrices) consist entirely of zeros, we just use 0; the meaning will be clear from the context.

Finally, we introduce a number of special matrices whose elements are interrelated in certain ways. These matrices are defined as follows:

Hermitian:	$A^* = A$
Skew-Hermitian:	$A^* = -A$
Symmetric (real):	$A^T = A$
Skew-symmetric (real):	$A^T = -A$
Unitary:	$A^*A = I$
Orthogonal (real):	$A^TA = I$

Exercises

1. Let A be the matrix

$$\begin{bmatrix} 1 & 0 & 0 & 0 \\ 0 & e^{i}/\sqrt{2} & 0 & e^{2i}/\sqrt{2} \\ 0 & 0 & 1 & 0 \\ 0 & -e^{-2i}/\sqrt{2} & 0 & e^{-i}/\sqrt{2} \end{bmatrix}$$

where $i = \sqrt{-1}$ and e is the exponential function. Write down the matrices $\bar{A}$, A^T, A^* and show that A is unitary.

2. Show that matrices which are (*a*) diagonal and (*b*) triangular (upper or lower) retain these properties on inversion. What can we say about tridiagonal and Hessenberg matrices?

3. Prove that

   (i) $(AB)^{-1} = B^{-1}A^{-1}$

   (ii) $(\mathbf{x}\mathbf{x}^*)^* = \mathbf{x}\mathbf{x}^*$

   (iii) $(A\mathbf{x})^* = \mathbf{x}^*A^*$

   (iv) $(AB)^* = B^*A^*$

2.2 Eigenvalues and eigenvectors

The basic problem with which we shall be concerned is the determination of the values of λ for which the n homogeneous linear equations in n unknowns

$$A\mathbf{x} = \lambda\mathbf{x} \tag{1}$$

have a solution $\mathbf{x}$ such that $x_i \neq 0$ for at least one i (a non-trivial solution). For this we require

$$\det(A - \lambda I) = 0 \tag{2}$$

and this is a polynomial equation of degree n in λ, called the *characteristic equation* of A. The left-hand side of equation (2) is called the *characteristic polynomial* of A. The n roots are called the *eigenvalues* of A, and these may or may not be distinct. If they are not distinct, then A is said to have multiple eigenvalues; an eigenvalue which occurs exactly m times is said to be of multiplicity m.

Corresponding to each value of λ satisfying equation (2) there is a non-trivial vector $\mathbf{x}$ satisfying equation (1) called the *eigenvector* of A corresponding to that eigenvalue. These eigenvectors are clearly arbitrary to the extent of a constant multiple, and it is convenient to choose this multiplier in such a way that $\mathbf{x}$ has some desirable property. Such vectors are then called normalized vectors, and the most useful normalization for our purposes is that $\mathbf{x}$ should satisfy

$$\mathbf{x}^*\mathbf{x} = 1.$$

If an $n \times n$ matrix A possesses n linearly independent eigenvectors, then these are said to form a *complete set.*

Theorem 2.1 If the matrix A has distinct eigenvalues, then there exists a complete set of linearly independent eigenvectors unique up to a constant multiple.

Proof Let the eigenvalues of A be $\lambda_1, \lambda_2, \ldots, \lambda_n$, and let the corresponding eigenvectors be $\mathbf{x}_1, \mathbf{x}_2, \ldots, \mathbf{x}_n$ respectively. Let m be the smallest number of linearly dependent eigenvectors. Then we must have $m \geqslant 2$, and, without loss of generality, we can write

$$\sum_{i=1}^{m} \alpha_i \mathbf{x}_i = \mathbf{0} \tag{3}$$

where $\alpha_1, \alpha_2, \ldots, \alpha_m$ are non-zero constants.

Premultiplying equation (3) by the matrix A, we have

$$\sum_{i=1}^{m} \alpha_i \lambda_i \mathbf{x}_i = \mathbf{0}, \tag{4}$$

and multiplying equation (3) by λ_m and subtracting equation (4) gives

$$\sum_{i=1}^{m-1} \alpha_i(\lambda_m - \lambda_i)\, \mathbf{x}_i = \mathbf{0}. \tag{5}$$

Since the eigenvalues are distinct, the coefficients of $\mathbf{x}_i$ are non-zero and this contradicts the definition of m. Thus equation (3) cannot hold for any $m \leqslant n$, and the linear independence is proved.

Now suppose there is a second eigenvector $\mathbf{y}_1$ corresponding to λ_1. Then we may write

$$\mathbf{y}_1 = \sum_{i=1}^{n} \beta_i \mathbf{x}_i \tag{6}$$

since the eigenvectors $\mathbf{x}_1, \ldots, \mathbf{x}_n$ are linearly independent and so a suitable linear combination represents any arbitrary vector of n elements. Multiplying equation (6) by A gives

$$\lambda_1 \mathbf{y}_1 = \sum_{i=1}^{n} \beta_i \lambda_i \mathbf{x}_i. \tag{7}$$

Multiplying equation (6) by λ_1 and subtracting from equation (7) then gives

$$\sum_{i=2}^{n} \beta_i(\lambda_i - \lambda_1)\, \mathbf{x}_i = 0,$$

which, from the independence of the $\mathbf{x}_i$ and the distinctness of the eigenvalues, implies that

$$\beta_i = 0, \quad i = 2, 3, \ldots, n.$$

Thus $\mathbf{y}_1$ is a multiple of $\mathbf{x}_1$ and the result is proved.

The structure of the system of eigenvectors of a matrix with multiple eigenvalues is far more complex. For example, the matrix

$$A = \begin{bmatrix} 2 & 3 \\ 0 & 2 \end{bmatrix}$$

has eigenvalues $\lambda_1 = \lambda_2 = 2$, with corresponding eigenvectors $\mathbf{x}_1 = \mathbf{x}_2 = \mathbf{e}_1$. Thus the matrix A has essentially only one eigenvector $\mathbf{e}_1$. On the other hand, the matrix

$$A = \begin{bmatrix} 2 & 0 \\ 0 & 2 \end{bmatrix}$$

again has eigenvalues $\lambda_1 = \lambda_2 = 2$, but corresponding to these we can find two independent eigenvectors $\mathbf{x}_1 = \mathbf{e}_1$ and $\mathbf{x}_2 = \mathbf{e}_2$. Clearly, any linear combination of $\mathbf{e}_1$ and $\mathbf{e}_2$ is also an eigenvector. Further discussion of the structure of the system of eigenvectors of a general matrix is left to Section 2.4.

Since the determinant of a matrix is equal to that of its transpose, it is easily seen that the eigenvalues of A^T are equal to those of A. Thus we have

$$A^T\mathbf{y} = \lambda\mathbf{y},$$

where $\mathbf{y}$ is an eigenvector of A^T corresponding to λ. It follows that

$$\mathbf{y}^T A = \lambda\mathbf{y}^T,$$

so that $\mathbf{y}^T$ is often called a left eigenvector of A corresponding to λ. The usual eigenvector $\mathbf{x}$ is then called the right eigenvector of A corresponding to λ.

Exercises

4. If A has n distinct eigenvalues, and X and Y are matrices formed by taking the right and left eigenvectors respectively as columns, show that we can take

$$X^T Y = I.$$

5. Consider how the eigenvectors of the matrix

$$A = \begin{bmatrix} a & 1 \\ 0 & b \end{bmatrix}$$

change as we let b tend to a.

6. Prove that the sum of the eigenvalues is the sum of the diagonal elements of A (the *trace* of A).

7. Prove that a matrix satisfies its own characteristic equation (the Cayley–Hamilton theorem).

2.3 Similarity transformations

Transformations of the matrix A of the form $R^{-1}AR$, where R is non-singular, are of fundamental importance from both the theoretical and the practical point of view, and are known as *similarity transformations*. The matrices A and $R^{-1}AR$ are said to be *similar*. Special importance is attached to similarity transformations where R is a unitary matrix, and in this case A and $R^{-1}AR$ are said to be *unitarily similar*. We return to this point in Section 2.9.

The usefulness of similarity transformations is a direct consequence of the result that the eigenvalues of a matrix are invariant under such transformations. For if

$$A\mathbf{x} = \lambda\mathbf{x}, \tag{8}$$

then it follows that

$$(R^{-1}AR)\, R^{-1}\mathbf{x} = \lambda R^{-1}\mathbf{x}. \tag{9}$$

Clearly the eigenvectors are premultiplied by R^{-1}.

Some of the most important methods which we shall present in subsequent chapters are based on the reduction of a matrix by a similarity transformation (or by a sequence of such transformations), to a matrix of special 'canonical' form, whose eigensolution (eigenvalues and eigenvectors) may be more easily obtained.

Let the matrix A have eigenvalues $\lambda_1, \lambda_2, \ldots, \lambda_n$ with corresponding eigenvectors $\mathbf{x}_1, \mathbf{x}_2, \ldots, \mathbf{x}_n$. Then

$$A\mathbf{x}_i = \lambda_i\mathbf{x}_i \qquad i = 1, 2, \ldots, n \tag{10}$$

and we may write this in matrix form as

$$AX = X\,\mathrm{diag}\,(\lambda_i), \tag{11}$$

where X is the $n \times n$ matrix whose ith column is $\mathbf{x}_i$.

If the eigenvectors are linearly independent, then X is non-singular, and so

$$X^{-1}AX = \mathrm{diag}\,(\lambda_i). \tag{12}$$

Thus we have shown that a matrix with a complete system of eigenvectors is similar to a diagonal matrix whose non-zero elements are the eigenvalues. A more general result is

Theorem 2.2 (*Schür's theorem*) *Any square matrix is unitarily similar to a triangular matrix with the eigenvalues on the diagonal.*

This result is best proved constructively using certain transformation matrices which are not introduced until the next chapter. The proof follows naturally from the results of Section 5.2, and is given there as an exercise.

Exercises

8. Show that a matrix with fewer than n linearly independent eigenvectors cannot be similar to a diagonal matrix.

9. Show that a Hermitian tridiagonal matrix may be transformed into a real symmetric tridiagonal matrix by a similarity transformation with a unitary diagonal matrix.

2.4 The Jordan canonical form

In view of exercise 8 of the previous section, a general matrix cannot be reduced to diagonal canonical form by similarity transformations, and we turn now to a consideration of the most compact form which such transformations produce. Although this form is of little practical importance, considerable insight can be gained into the structure of the system of eigenvectors.

We begin by defining matrices $J_r(\lambda)$ by the relations

$$J_1(\lambda) = [\lambda],$$

$$J_r(\lambda) = \begin{bmatrix} \lambda & 1 & & & & \\ & \lambda & 1 & & & \\ & & \cdot & \cdot & & \\ & & & \cdot & \cdot & \\ & & & & \cdot & \cdot \\ & & & & \lambda & 1 \\ & & & & & \lambda \end{bmatrix}, \quad r > 1$$

where $J_r(\lambda)$ is an $r \times r$ matrix with an eigenvalue λ of multiplicity r, but only one eigenvector $\mathbf{x} = \mathbf{e}_1$. The matrix $J_r(\lambda)$ is called a *simple Jordan submatrix* of order r.

Now let A be a matrix of order n with s distinct eigenvalues $\lambda_1, \lambda_2, \ldots, \lambda_s$ of multiplicities $m_1, m_2, \ldots, m_s$ so that

$$\sum_{i=1}^{s} m_i = n.$$

Then we have

Theorem 2.3 There exists a non-singular matrix R such that $R^{-1}AR$ has simple Jordan submatrices $J_r(\lambda_i)$ isolated along the diagonal with all other elements equal to zero. If there are p submatrices of orders $r_j, j = 1, 2, \ldots, p$ associated with any λ_i, then

$$\sum_{j=1}^{p} r_j = m_i.$$

The matrix $R^{-1}AR$ is called the *Jordan canonical form* of A, and is unique apart from the ordering of the submatrices along the diagonal. It is the most compact form to which a general matrix may be reduced by a similarity transformation. If the matrix has a complete system of eigenvectors, we have already seen that it can be reduced to diagonal form, and thus all the Jordan submatrices are of order unity.

The total number of independent eigenvectors is equal to the number of submatrices in the Jordan canonical form. For example, a matrix A of order eight which can be reduced to the form

$$J = R^{-1}AR = \begin{bmatrix} \lambda_1 & 1 & & & & & & \\ & \lambda_1 & 1 & & & & & \\ & & \lambda_1 & & & & & \\ & & & \lambda_1 & 1 & & & \\ & & & & \lambda_1 & & & \\ & & & & & \lambda_2 & 1 & \\ & & & & & & \lambda_2 & \\ & & & & & & & \lambda_3 \end{bmatrix}$$

$$= \begin{bmatrix} J_3(\lambda_1) & & & \\ & J_2(\lambda_1) & & \\ & & J_2(\lambda_2) & \\ & & & J_1(\lambda_3) \end{bmatrix} \tag{13}$$

has four independent eigenvectors, two corresponding to λ_1, and one each corresponding to λ_2 and λ_3. Further, the eigenvectors of A are Re_1, Re_4, Re_6 and Re_8. Clearly, there can be at most m linearly independent eigenvectors corresponding to an eigenvalue of multiplicity m.

Corresponding to the matrix J of equation (13) we have

$$J-\lambda I = \begin{bmatrix} J_3(\lambda_1-\lambda) & & & \\ & J_2(\lambda_1-\lambda) & & \\ & & J_2(\lambda_2-\lambda) & \\ & & & J_1(\lambda_3-\lambda) \end{bmatrix}. \tag{14}$$

The determinants

$$\det(J_r(\lambda_i-\lambda))$$

are known as the *elementary divisors* of A. Thus the elementary divisors corresponding to equation (14) are $(\lambda_1-\lambda)^3$, $(\lambda_1-\lambda)^2$, $(\lambda_2-\lambda)^2$ and $(\lambda_3-\lambda)$.

A matrix with distinct eigenvalues always has elementary divisors which are polynomials of degree one in λ (*linear* elementary divisors). If the eigenvalues are not distinct, the elementary divisors may or may not be linear. If at least one is non-linear, then the matrix has fewer than n linearly independent eigenvectors, and is said to be *defective*.

Exercises

10. Prove that the eigenvectors of the matrix J given by equation (13) are in fact $\mathbf{e}_1$, $\mathbf{e}_4$, $\mathbf{e}_6$ and $\mathbf{e}_8$.

11. If $AB = BA$ (i.e. A and B *commute*), and if, further, both have linear elementary divisors, show that they share a common system of eigenvectors. What can we say if (*a*) A has, (*b*) A and B have at least one non-linear elementary divisor?

2.5 Some properties of Hermitian matrices

Matrices of Hermitian type occur frequently in applications. Such matrices have a number of important special properties which are very useful for computational purposes, and some of those are now summarized. The results apply directly to symmetric matrices, if the elements are real.

It is convenient to introduce the following definitions, which will be required later:

A Hermitian matrix A is

$$\left.\begin{cases}\textit{positive definite}\\ \textit{non-negative definite}\\ \textit{negative definite}\\ \textit{non-positive definite}\end{cases}\right\} \text{ if } \mathbf{x}^*A\mathbf{x} \left\{\begin{matrix} > \\ \geqslant \\ < \\ \leqslant \end{matrix}\right\} 0$$

for all non-trivial vectors $\mathbf{x}$.

Now suppose that the Hermitian matrix A satisfies

$$A\mathbf{x} = \lambda\mathbf{x}. \tag{15}$$

Then

$$\mathbf{x}^*A\mathbf{x} = \lambda\mathbf{x}^*\mathbf{x}. \tag{16}$$

Since $(\mathbf{x}^*A\mathbf{x})^* = \mathbf{x}^*A\mathbf{x}$ is real, and since $\mathbf{x}^*\mathbf{x}$ is real, it follows that λ is real. Thus we have proved the result that the eigenvalues of a Hermitian matrix are real.

If a Hermitian matrix has distinct eigenvalues, then it can readily be shown that the eigenvectors $\mathbf{x}_i$ satisfy

$$\mathbf{x}_i^*\mathbf{x}_j = 0 \quad i \neq j \tag{17}$$

and if we normalize so that

$$\mathbf{x}_i^*\mathbf{x}_i = 1 \tag{18}$$

then the matrix X whose columns are the eigenvectors $\mathbf{x}_i$ satisfies

$$X^*X = I \tag{19}$$

and is therefore a unitary matrix. The eigenvectors are said to form an *orthonormal* set.

Thus, using equations (10) and (11), it follows that if A is a Hermitian matrix, with distinct eigenvalues, there exists a unitary matrix X such that

$$X^*AX = \operatorname{diag}(\lambda_i), \quad \lambda_i \text{ real.} \tag{20}$$

An extremely important generalization of this result is that a unitary matrix exists which transforms a Hermitian matrix A to diagonal form *even if A has multiple eigenvalues*. This result is an immediate consequence of Theorem 2.2. It follows that the elementary divisors of a Hermitian matrix are linear, and a Hermitian matrix cannot be defective.

Exercises

12. Derive equation (19).

13. Use Theorem 2.2 to show that a Hermitian matrix is unitarily similar to a diagonal matrix.

14. Prove that a Hermitian matrix A is

$$\left\{\begin{array}{l}\text{positive definite}\\ \text{non-negative definite}\\ \text{negative definite}\\ \text{non-positive definite}\end{array}\right\} \text{ if and only if the eigenvalues are } \left\{\begin{array}{l}\text{positive}\\ \text{non-negative}\\ \text{negative}\\ \text{non-positive}\end{array}\right\}.$$

2.6 Vector and matrix norms

It is useful to have some measure of the 'size' of a vector or matrix, and to this end we introduce the concept of a *norm*. The norm of a vector $\mathbf{x}$ is denoted by $\|\mathbf{x}\|$ and satisfies the following relations:

(i) $\|\mathbf{x}\| > 0$ unless $\mathbf{x} = 0$,
(ii) $\|k\mathbf{x}\| = |k| \, \|\mathbf{x}\|$ where k is a (complex) scalar,
(iii) $\|\mathbf{x}+\mathbf{y}\| \leqslant \|\mathbf{x}\| + \|\mathbf{y}\|$.

We shall require only two simple norms, the cases $p = 2$ and $p = \infty$ of the general L_p norm defined for $p \geqslant 1$ by

$$\|\mathbf{x}\|_p = \left(\sum_{i=1}^{n} |x_i|^p\right)^{1/p}, \tag{21}$$

where we interpret $\|\mathbf{x}\|_\infty$ as the maximum value of $|x_i|$ over all i, i.e. $\max_i |x_i|$. $\|\mathbf{x}\|_2$ is normally referred to as the *Euclidean length* of a vector.

In a similar fashion, the norm of a matrix A is denoted by $\|A\|$ and satisfies the relations:

(i) $\|A\| > 0$ unless $A = 0$,
(ii) $\|kA\| = |k|\,\|A\|$ where k is a (complex) scalar,
(iii) $\|A+B\| \leqslant \|A\|+\|B\|$,
(iv) $\|AB\| \leqslant \|A\|\,\|B\|$.

For any vector norm, we can define a corresponding matrix norm

$$\|A\| = \max_{\mathbf{x}\neq 0} \frac{\|A\mathbf{x}\|}{\|\mathbf{x}\|} = \max_{\|\mathbf{x}\|=1} \|A\mathbf{x}\|. \tag{22}$$

It can readily be verified that this definition satisfies the above conditions for a matrix norm, and this norm is called the matrix norm *subordinate* to the particular vector norm.

Clearly, the subordinate matrix norm satisfies

$$\|A\mathbf{x}\| \leqslant \|A\|\,\|\mathbf{x}\|; \tag{23}$$

matrix and vector norms for which (23) are satisfied are said to be *compatible.*

The matrix norms subordinate to the L_2 and L_∞ vector norms are

$$\|A\|_2 = (\text{maximum eigenvalue of } A^*A)^{1/2} \tag{24}$$

$$\|A\|_\infty = \max_i \sum_{j=1}^{n} |a_{ij}|. \tag{25}$$

The norm defined by equation (24) is often called the *spectral norm.*

Another important norm compatible with the L_2 vector norm is the Euclidean or Schür norm defined by

$$\|A\|_{\mathrm{E}} = \left(\sum_{i,j} |a_{ij}|^2\right)^{1/2}. \tag{26}$$

This norm is useful because it can readily be computed.

Exercises

15. Derive equations (24) and (25).

16. Show that the Euclidean matrix norm is not subordinate to any vector norm.

17. Show that the following invariance properties hold:

(i) $\|R\mathbf{x}\|_2 = \|\mathbf{x}\|_2$

(ii) $\|RA\|_2 = \|A\|_2$

(iii) $\|RAR^*\|_2 = \|A\|_2$

(iv) $\|RA\|_{\mathrm{E}} = \|A\|_{\mathrm{E}}$,

where R is unitary, and x and A are arbitrary.

2.7 Theorems on bounds for the eigenvalues

It is often convenient, from both the theoretical and the practical point of view, to locate the eigenvalues of a given matrix in bounded regions of the complex plane. Information of this type is useful in the subsequent application of a number of iterative methods for obtaining more precise eigenvalues, and also plays an important role in an analysis of the changes in the eigenvalues brought about by introducing perturbations in the elements of the matrix (see Section 2.8). A fundamental result is the following theorem, often called the Gershgorin circle theorem.

Theorem 2.4 *Let A be an $n \times n$ matrix, and let C_i, $i = 1, 2, \ldots, n$ be the discs with centres a_{ii} and radii $R_i = \sum_{\substack{k=1\\k\neq i}}^{n} |a_{ik}|$. Let D denote the union of the discs C_i. Then all the eigenvalues of A lie within D.*

Proof Let λ be an eigenvalue of A, and $\mathbf{x}$ a corresponding eigenvector normalized so that $\max_i |x_i| = 1$. Then

$$\lambda \mathbf{x} = A\mathbf{x},$$

i.e.

$$(\lambda - a_{ii})\, x_i = \sum_{\substack{k=1\\k\neq i}}^{n} a_{ik} x_k, \quad i = 1, 2, \ldots, n.$$

Now if $|x_r| = 1$, then

$$|\lambda - a_{rr}| \leqslant \sum_{\substack{k=1\\k\neq r}}^{n} |a_{rk}|\, |x_k|$$

$$\leqslant \sum_{\substack{k=1\\k\neq r}}^{n} |a_{rk}| = R_r.$$

Thus, the eigenvalue λ lies in the disc C_r. Since λ is arbitrary, it follows that all the eigenvalues of A must lie in the union of the discs, i.e. in D.

Another basic result is

Theorem 2.5 *Let A be a real symmetric matrix, and let $\mathbf{x}$ be an arbitrary real or complex vector. Then*

(i) $$\lambda_1 = \max_{\mathbf{x}\neq 0} \frac{\mathbf{x}^* A \mathbf{x}}{\mathbf{x}^* \mathbf{x}}$$

(ii) $$\lambda_n = \min_{\mathbf{x}\neq 0} \frac{\mathbf{x}^* A \mathbf{x}}{\mathbf{x}^* \mathbf{x}}$$

where the eigenvalues are ordered so that

$$\lambda_1 \geqslant \lambda_2 \geqslant \ldots \geqslant \lambda_n.$$

Proof

(i) For any vector $\mathbf{x}$

$$\frac{\mathbf{x}^*A\mathbf{x}}{\mathbf{x}^*\mathbf{x}} - \lambda_1 = \frac{\mathbf{x}^*(A-\lambda_1 I)\mathbf{x}}{\mathbf{x}^*\mathbf{x}} \leqslant 0,$$

where we have used the result of exercise 14 of Section 2.5. Clearly equality holds when $\mathbf{x}$ is the eigenvector corresponding to λ_1.

(ii) is proved similarly.

A number of other results are left as exercises for the reader.

Exercises

18. Let $\rho(A) = \max_i |\lambda_i|$ (the *spectral radius* of A). Show that

(*a*)
$$\|A\| \geqslant \rho(A),$$

for any norm,

(*b*)
$$\rho(A) \leqslant \max_r\{|a_{rr}| + R_r\}$$
$$\{\rho(A^{-1})\}^{-1} \geqslant \min_r\{|a_{rr}| - R_r\}$$

in the notation of Theorem 2.4, where A is non-singular,

(*c*)
$$\{\rho(A^*A)^{-1}\}^{-1} \leqslant |\lambda_i|^2 \leqslant \rho(A^*A),$$

for any i, where A is non-singular.

19. Prove that

$$\sum_{i=1}^{n} |\lambda_i|^2 \leqslant \|A\|_E^2.$$

(Hint: use the result (Theorem 2.2) that there exists a unitary matrix R such that $R^*AR = T$, where T is triangular with the eigenvalues of A on the diagonal.)

2.8 Condition of the eigenvalue problem

We now consider the effects on the eigenvalues of a matrix A of making small changes to the elements of A. Some of the eigenvalues may be very sensitive to such changes, while others may be insensitive. If small perturbations in the elements of A can lead to arbitrarily large perturbations in some of the eigenvalues, then the eigenvalue problem for those eigenvalues is said to be *ill-conditioned.* The significance of this is that in computation, rounding errors are inevitably introduced, and it is therefore difficult to obtain accurate results for an ill-conditioned problem.

We consider first the case where A is Hermitian. Then if B represents a matrix of perturbations of A which is also assumed to be Hermitian, and $|b_{ij}| \leqslant \epsilon$, it is possible to show (Wilkinson, 1965) that

$$\left\{\sum_{i=1}^{n} (\delta\lambda_i)^2\right\}^{1/2} \leqslant n\epsilon, \tag{27}$$

where $\delta\lambda_i$ represents the perturbation of λ_i brought about by working with the matrix $A+B$ instead of A. Thus it follows that the eigenvalue problem for a Hermitian matrix is always well-conditioned.

For non-Hermitian matrices, the situation is far more complex, although we have the result (Wilkinson, 1965)

$$|\delta\lambda_i| \leqslant \|B\|_2/|\mathbf{y}_i^T\mathbf{x}_i|, \tag{28}$$

where $\mathbf{x}_i$ and $\mathbf{y}_i$ are respectively the right- and left-hand eigenvectors of A corresponding to a simple (i.e. of multiplicity one) eigenvalue λ_i normalized so that their Euclidean length is unity. If $\mathbf{x}_i$ and $\mathbf{y}_i$ are real, then $\mathbf{y}_i^T\mathbf{x}_i$ represents the cosine of the angle θ between them, and this can be arbitrarily small.

For example, consider the matrix

$$\begin{bmatrix} a & 1 \\ 0 & a+\epsilon \end{bmatrix}.$$

The eigenvalues are $\lambda_1 = a$ and $\lambda_2 = a+\epsilon$, with corresponding eigenvectors

$$\mathbf{x}_1 = (1, 0)^T, \quad \mathbf{y}_1 = \frac{(-\epsilon, 1)^T}{\sqrt{1+\epsilon^2}}$$

$$\mathbf{x}_2 = \frac{(1, \epsilon)^T}{\sqrt{1+\epsilon^2}}, \quad \mathbf{y}_2 = (0, 1)^T.$$

Thus

$$\mathbf{x}_1^T\mathbf{y}_1 = \frac{-\epsilon}{\sqrt{1+\epsilon^2}}, \quad \mathbf{x}_2^T\mathbf{y}_2 = \frac{\epsilon}{\sqrt{1+\epsilon^2}},$$

and both these quantities tend to zero with ϵ. When $\epsilon = 0$, then the matrix is defective. The problem of finding the eigenvalues corresponding to non-linear elementary divisors will normally be ill-conditioned, although it is worth mentioning that matrices having exact non-linear elementary divisors are extremely rare in practice. We note in passing that for Hermitian matrices, $\mathbf{y}_i^T\mathbf{x}_i = 1$, and so we have confirmation of the result given above that the eigenvalue problem for a Hermitian matrix is well-conditioned.

An analysis of the sensitivity of the eigenvectors of a matrix with respect to perturbations of the elements is extremely complex. We merely point out that even in the Hermitian case, eigenvectors corresponding to close eigenvalues are very sensitive.

2.9 Stability of similarity transformation methods

The condition referred to above is purely a function of the problem, and thus if a problem is ill-conditioned, then we should expect difficulty with any numerical method. However, good methods should be such that their application does not cause the condition to deteriorate, and does not transform a well-conditioned problem into an ill-conditioned one. We can consider such methods to be *stable*. In particular, algorithms based on these methods should always provide accurate results when used to solve the eigenvalue problem for a Hermitian matrix.

The basis for a number of popular methods for solving the eigenvalue problem is the use of similarity transformations. Suppose that $R^{-1}AR$ is obtained from A. Then if $\mathbf{x}_i$ and $\mathbf{y}_i$ are right- and left-hand eigenvectors of A corresponding to λ_i, the corresponding right- and left-hand eigenvectors of $R^{-1}AR$ are $R^{-1}\mathbf{x}_i$ and $R^T\mathbf{y}_i$. Further, if R is a unitary matrix, these vectors have the same Euclidean lengths as $\mathbf{x}_i$ and $\mathbf{y}_i$ (exercise 17, Section 2.6). In particular, the quantity $\cos\theta_i$, where θ_i is the angle between the vectors $\mathbf{y}_i$ and $\mathbf{x}_i$, is preserved by unitary transformations, and equation (28) shows that *the condition of the eigenvalue problem cannot deteriorate*. This is a very important observation.

If R is not unitary, then the Euclidean lengths of $R^{-1}\mathbf{x}_i$ and $R^T\mathbf{y}_i$ may be much greater than those of $\mathbf{x}_i$ and $\mathbf{y}_i$ respectively, permitting the possibility of serious deterioration in the condition. However it is possible to use non-unitary transformation matrices in a satisfactory way by using a device known as 'pivoting', which enables the elements of the transforming matrix to be bounded (see also Section 3.5). This process is known as stabilization, and in practice can lead to algorithms which are almost as stable as those based on unitary transformations, and which are usually much simpler.

3

Reductions and transformations

3.1 Introduction

In Chapter 2, it was shown that the eigensystem of a similarity transformation of a general matrix A was related in a straightforward manner to the eigensystem of A itself. To be more precise, for an arbitrary non-singular matrix H the eigenvalues of the matrix $H^{-1}AH$ are the same as those of the matrix A and the eigenvectors of the matrix $H^{-1}AH$ are given by the vectors $H^{-1}\mathbf{x}$ where $\mathbf{x}$ is an eigenvector of the matrix A. The property of the invariance of the eigenvalues is most appealing as it suggests that it might be possible to find the eigenvalues of a matrix A by computing those of a related matrix $H^{-1}AH$ of simpler form. The basic differences in the algorithms of this type to be studied in subsequent chapters are defined almost entirely in terms of the structures of the simpler forms of the matrix. Thus we shall see that the Jacobi algorithm aims at providing a sequence of similarity transformations which will reduce the Hermitian matrix A to diagonal form. The techniques of Givens and Householder provide reductions to tridiagonal form if the matrix A is Hermitian, and to Hessenberg form if A is a general matrix.

Rather than compute a single transforming matrix H we proceed in an iterative fashion gradually reducing the original matrix to one of a simpler form. In this chapter, we will introduce and study the basic types of transformation which will form the building blocks of our subsequent algorithms. We shall also see how these transformation matrices can be employed to provide a decomposition of a matrix into the product of two matrices of particular type. These fundamental transformation matrices are generally referred to as *elementary matrices.*

3.2 Elementary operation matrices

Our first class of elementary matrices consists of those matrices whose effect is one of the following:

(i) multiplication of a particular row or column of a matrix by a constant k

(ii) addition of a scalar multiple of one row or column to another row or column

(iii) interchange of two rows, or of two columns.

The following notation will be employed for the above three types of matrices:

(i) $$M_i = \text{diag}\,(1, \ldots, 1, k, 1, \ldots, 1)$$

where the constant k is the ith diagonal entry

(ii)
$$S_{ji} = \begin{bmatrix} 1 & & & & & & \\ & 1 & & & & & \\ & & \ddots & & & & \\ & & & 1 & & & \\ & & & \vdots & \ddots & & \\ & & & k & & 1 & & \\ & & & & & & \ddots & \\ & & & & & & & 1 \end{bmatrix}$$

where S_{ji} is the unit matrix except for an entry k in the (j, i) position.

(iii) I_{ij} is the identity matrix with the modifications that the (i, i) and (j, j) diagonal entries are zero, and the (i, j) and (j, i) entries are one.

Premultiplication (postmultiplication) of the matrix A by the elementary matrix M_i causes the entries in the ith row (column) of the matrix A to be multiplied by the constant k. Likewise premultiplication (postmultiplication) of the matrix A by S_{ji} adds k times row i (column j) to row j (column i) of the matrix A. Finally premultiplication (postmultiplication) of the matrix A by I_{ij} causes the interchange of rows (columns) i and j of the matrix A.

The above three matrices will be of use in later sections and in particular in the solution of linear systems.

Exercises

1. If we write $M_i(k)$, $S_{ij}(k)$ to show the dependence of these matrices on the non-zero parameter k, prove that

$$\{M_i(k)\}^{-1} = \left\{M_i\left(\frac{1}{k}\right)\right\}$$
$$\{S_{ij}(k)\}^{-1} = \{S_{ij}(-k)\}$$
$$\{I_{ij}\}^{-1} = I_{ij}.$$

2. Let the matrix L_i be the unit matrix except for the ith column which is given by

$$(0, 0, \ldots, 1, -r_{i+1,i}, -r_{i+2,i}, \ldots, -r_{n,i})^T.$$

Show that L_i may be written as a product of matrices of the type $S_{ji}(k_{ji})$. (L_i is an *elementary lower triangular matrix*.)

3. Verify that the product $L_1L_2 \ldots L_{n-1}$ is a lower triangular matrix with unit diagonal (i.e. diagonal elements unity) and with $-r_{i,j}$ in the (i, j) position for $j<i$. Compare this product with the product $L_{n-1}L_{n-2} \ldots L_1$.

4. A *permutation matrix* is defined to be a product of matrices of the form I_{ij}. Show by induction that a permutation matrix has precisely one non-zero element, equal to one, in each row and column.

5. Show that premultiplication by the matrix L_i of exercise 2 has the effect of subtracting $r_{i,j}$ times the ith row from the jth row, $j>i$.

3.3 Elementary unitary matrices

The second type of fundamental matrix which we shall require is known as an *elementary unitary matrix*. This matrix, designated $R(j, k)$ or more briefly when no confusion can arise, as R, is defined to be the unit matrix subject to the following alterations:

$$\begin{aligned} r_{jj} &= e^{i\alpha} \cos \theta & r_{jk} &= e^{i\beta} \sin \theta \\ r_{kj} &= -e^{-i\beta} \sin \theta & r_{kk} &= e^{-i\alpha} \cos \theta \end{aligned} \tag{1}$$

where α, β, θ are real and $i = \sqrt{-1}$. More specifically the above matrix is referred to as a *plane rotation* in the plane (j, k). It may easily be verified that the matrix R is indeed unitary so R^{-1} is simply R^*. By inspection we see that R^* is itself a plane rotation corresponding to the quantities α_*, β_* where

$$\alpha_* = -\alpha, \quad \beta_* = \beta+\pi.$$

It will in fact frequently be unnecessary for us to consider plane rotations of such generality. In particular, the choice $\alpha = \beta = 0$ leads to a real matrix R which is orthogonal.

Premultiplication (postmultiplication) of a matrix A by the matrix R affects only rows (columns) j, k of the matrix A. It follows that the only elements modified in the similarity transformation RAR^* of A lie in the jth and kth rows and columns.

Exercises

6. Verify that the eigenvalues of the matrix R lie on the unit circle (i.e. have modulus equal to one).

7. Consider the more general definition of the four modified elements of the matrix R, namely

$$\begin{aligned} r_{jj} &= e^{i\alpha} \cos \theta & r_{jk} &= e^{i\beta} \sin \theta \\ r_{kj} &= e^{i\gamma} \sin \theta & r_{kk} &= e^{i\delta} \cos \theta \end{aligned}$$

where α, β, γ, δ, θ are all real. Find a condition on the parameters α, β, γ, δ which will ensure that R is unitary.

8. Show that the matrix R in the text may be written in terms of a product of elementary operation matrices provided $\cos\theta \neq 0$.

3.4 Elementary Hermitian matrices

The third class of elementary matrices we require consists of those of the form

$$P_\mu = I - \mu\mathbf{w}\mathbf{w}^*,$$

where $\mathbf{w}^*\mathbf{w} = 1$, and μ is a complex constant. It is left as an exercise to prove that P_μ is a unitary matrix if

$$\mu + \bar{\mu} = \mu\bar{\mu},$$

and is Hermitian if $\mu = 2$.

Given the distinct vectors $\mathbf{x}$ and $\mathbf{y}$ satisfying the condition

$$\mathbf{x}^*\mathbf{x} = \mathbf{y}^*\mathbf{y}$$

we can choose μ and $\mathbf{w}$ so that

$$P_\mu\mathbf{x} = \mathbf{y}.$$

If P_μ is required to be Hermitian ($\mu = 2$) then an additional restriction is that $\mathbf{x}^*\mathbf{y}$ be real. Using the definition of P_μ we have that

$$\mathbf{y} = P_\mu\mathbf{x} = \mathbf{x} - \mu(\mathbf{w}^*\mathbf{x})\,\mathbf{w}$$

and thus

$$\mathbf{w} = \frac{1}{\alpha}(\mathbf{x}-\mathbf{y})$$

where

$$|\alpha|^2 = (\mathbf{x}-\mathbf{y})^*(\mathbf{x}-\mathbf{y}).$$

Further $\alpha = \mu\mathbf{w}^*\mathbf{x}$, so the complex constant μ is given by

$$\mu = \frac{\alpha\bar{\alpha}}{(\mathbf{x}-\mathbf{y})^*\,\mathbf{x}} = \frac{(\mathbf{x}-\mathbf{y})^*(\mathbf{x}-\mathbf{y})}{(\mathbf{x}-\mathbf{y})^*\,\mathbf{x}}.$$

In later sections of this and subsequent chapters, we will make particular choices of the form of the vector $\mathbf{y}$. The exercises for this section include some examples of this type.

Exercises

9. Derive the conditions on the complex constant μ for the matrix $P_\mu = I - \mu\mathbf{w}\mathbf{w}^*$, where $\mathbf{w}^*\mathbf{w} = 1$, to be (*a*) unitary, (*b*) Hermitian.

10. For an arbitrary vector $\mathbf{x} = (x_1, x_2, \ldots, x_n)^T$ show that it is possible to choose P_2 so that $\mathbf{y}$ has the form $(x_1, x_2, \ldots, x_r, y_{r+1}, 0, 0, \ldots, 0)^T$.

11. For an arbitrary vector $\mathbf{x}$ show that it is possible to choose μ and $\mathbf{w}$ in P_μ such that $\mathbf{y}$ is of the form $k\mathbf{e}_r$ where k is a real positive constant and $\mathbf{e}_r$ is the rth unit coordinate vector.

12. If in question 11 we restrict μ to $\mu = 2$, show that we may still determine a $\mathbf{w}$ such that $\mathbf{y} = k\mathbf{e}_r$ but that k will only be real if x_r is real.

3.5 Gaussian elimination

The solution of the set of linear equations

$$A\mathbf{x} = \mathbf{b}$$

where A is a dense matrix, and $\mathbf{b}$ is a known vector, is a problem which occurs frequently in numerical analysis, science and technology. We do not intend to cover, in this text, the many techniques for finding the solution vector $\mathbf{x}$. We will content ourselves with a straightforward development of the most frequently used algorithm, known as *Gaussian elimination*. This is included for two reasons: first for reference purposes in connection with later chapters and second as an example of the use of the elementary operation matrices. A fuller and more rigorous treatment may be found in the books by Wilkinson (1965) and by Forsythe and Moler (1967).

In exercise 2 of Section 3.2, the matrix L_i was introduced. It is the unit matrix except for the ith column which is given by

$$(0, 0, \ldots, 1, -r_{i+1,i}, -r_{i+2,i}, \ldots, -r_{n,i})^T.$$

In exercise 5, premultiplication by the matrix L_i was shown to have the effect of subtracting $r_{i,j}$ times the ith row from the jth row of the matrix A for $j > i$. Using these matrices it is possible to decompose the matrix A into the form LU where L is a lower triangular matrix with unit diagonal and U is an upper triangular matrix.

Suppose the element a_{11} of the matrix A is non-zero. Then premultiplication of A by the matrix L_1 modifies rows 2 to n of the matrix A by adding to them $(-r_{j1})$ $(j = 2, \ldots, n)$ times row one. If we choose $r_{j1} = a_{1j}/a_{11}$ it follows that the matrix $A_1 = L_1 A$ has zeros in positions 2 to n of the first column. If $a_{22}^{(1)}$, the (2, 2) element of A_1, is non-zero we may apply a similar technique to reduce elements 3 to n of the second column of A_1 to zero. If all the *pivot elements* $a_{j+1,j+1}^{(j)}$ of $A_j = L_j L_{j-1} \ldots L_1 A$ are non-zero for $j = 0, \ldots, n-2$ the above process is well-defined. If however at some stage we find that $a_{j+1,j+1}^{(j)} = 0$ then it is no longer possible to proceed with the decomposition in the above manner. Nevertheless, if one of the elements $a_{i,j+1}^{(j)}$, $i = j+2, \ldots, n$, is non-zero we may interchange rows i and $j+1$ to obtain a non-zero pivot. The interchanging of these two rows is, in terms of our elementary operation matrices, equivalent to premultiplying by $I_{j+1,i}$. If it is not possible to find a non-zero pivot then the matrix A is singular.

In practice, one does not only interchange rows when a zero pivot is encountered. Instead most programs available in the literature change pivots in the above manner to ensure that all the elements r_{ij} of the matrices L_i have modulus less than or equal to one. This process, which is known as *partial pivoting*, effectively stabilizes the basic Gaussian elimination procedure.

Ignoring the possibility of pivoting we see that the above process takes the form

$$L_{n-1}L_{n-2} \ldots L_1 A = U,$$

where U is an upper triangular matrix. The inverse of the matrix L_i is of the same form as L_i with the signs of the off-diagonal elements changed. Moreover (see exercise 3, Section 3.2) the product $L_{n-1} \ldots L_1$ is a complicated function of the elements of the L_i matrices, although a product of the form $L_1 \ldots L_{n-1}$ can be easily computed. It follows that we may easily store the above decomposition in the form

$$A = (L_1^{-1}L_2^{-1} \ldots L_{n-1}^{-1})\, U = LU$$

where L is a lower triangular matrix with unit diagonal. A more detailed discussion of the above algorithm and of its refinements may be found in the book of Forsythe and Moler (1967).

Exercises

13. Show that it is possible to pivot by columns instead of by rows.

14. Discuss the possibility of the decomposition $A = UL$.

15. Show that for a non-singular matrix A, the LU decomposition (with the diagonal of L equal to the unit matrix) is unique.
(Hint: suppose there exist two such decompositions

$$A = L_1U_1 = L_2U_2.$$

Prove that L_1, L_2, U_1, U_2 are all non-singular and then consider the implications of the equation

$$(L_1)^{-1} L_2 = U_1(U_2)^{-1}.)$$

16. If A is symmetric and positive definite then it may be shown that no pivots are zero. By appealing to the uniqueness of the LU decomposition as defined above, show that there exists a lower triangular matrix G such that

$$A = GG^T.$$

(This is the *Cholesky decomposition* of a positive definite matrix and can be performed in an extremely stable manner without the need for pivoting.)

3.6 Unitary decomposition of a matrix

The solution of the linear system $A\mathbf{x} = \mathbf{b}$ is usually carried out by means of the LU decomposition process outlined in the previous section. However one of the drawbacks of such a technique is the need to carry out interchanges of rows. On certain occasions it is either not desirable or not possible to proceed in such a manner.

In this section, we therefore turn to an alternative process based on the use of the elementary unitary matrices introduced in an earlier section. Our aim will be to decompose the matrix A into a product of a unitary matrix Q and an upper triangular matrix U. We study first the procedure based on the use of elementary unitary matrices. If we premultiply the vector $\mathbf{x}$ by the rotation matrix $R(j, k)$ then only elements y_j and y_k of the vector $\mathbf{y} = R(j, k)\,\mathbf{x}$ differ from the corresponding components of $\mathbf{x}$ (that is $y_p = x_p, p \neq j, k$). In fact,

$$\begin{aligned} y_j &= r_{jj}x_j + r_{jk}x_k \\ &= \mathrm{e}^{i\alpha}\cos\theta\, x_j + \mathrm{e}^{i\beta}\sin\theta\, x_k \end{aligned}$$

$$\begin{aligned} y_k &= r_{kj}x_j + r_{kk}x_k \\ &= -\mathrm{e}^{-i\beta}\sin\theta\, x_j + \mathrm{e}^{-i\alpha}\cos\theta\, x_k. \end{aligned}$$

We may choose the parameters α, β, θ in order to make y_k (or y_j) zero.

It follows that we may annihilate a single entry in the matrix A by premultiplication by a suitable $R(j, k)$. However our aim is to annihilate *all* the elements below the diagonal of the matrix A. It is important that we order our choice of planes of rotation in order to accomplish this. The standard procedure is to reduce to zero all the elements lying below the diagonal in a particular column, before proceeding to the next column. To annihilate the first column we successively make the elements in positions (2, 1), (3, 1), . . . , $(n, 1)$ vanish by rotations in the planes (1, 2), (1, 3), . . . , $(1, n)$. Then proceeding to the second column we annihilate the elements (3, 2), (4, 2), . . . , $(n, 2)$ by rotations in the planes (2, 3), (2, 4), . . . , $(2, n)$ and so on. It may be easily verified that this strategy does not make any element which has been forced to vanish non-zero at a later stage. It follows that we may reduce our matrix A to upper triangular form by, at most, $\frac{1}{2}n(n-1)$ plane rotations. It is worth bearing in mind that it is not necessary to carry out a rotation if the element to be annihilated is already zero.

Example Illustrate the above ideas by carrying through the decomposition for the following 3×3 *matrix*

$$A_0 = \begin{bmatrix} 12 & -20 & 41 \\ 9 & -15 & -63 \\ 20 & 50 & 35 \end{bmatrix}.$$

The first step is the annihilation of the element in the (2, 1) position by a rotation in the (1, 2) plane of the form (since A_0 real implies $\alpha = \beta = 0$)

$$R(1,2) = \begin{bmatrix} \cos\theta & \sin\theta & 0 \\ -\sin\theta & \cos\theta & 0 \\ 0 & 0 & 1 \end{bmatrix}.$$

The angle of rotation θ is chosen to make the (2, 1) element of the matrix $A_1 = R(1,2)\,A_0$ zero. From the equations of this section we see that the appropriate choice is $\cos\theta = \frac{4}{5}$, $\sin\theta = \frac{3}{5}$. Then

$$A_1 = \begin{bmatrix} 15 & -25 & -5 \\ 0 & 0 & -75 \\ 20 & 50 & 35 \end{bmatrix}.$$

The next step is a rotation $R(1, 3)$ to annihilate the element in the (3, 1) position of $R(1,3)\,A_1$. The appropriate rotation angle is now given by $\cos\theta = \frac{3}{5}$, $\sin\theta = \frac{4}{5}$ and

$$A_2 = \begin{bmatrix} 3/5 & 0 & 4/5 \\ 0 & 1 & 0 \\ -4/5 & 0 & 3/5 \end{bmatrix} \begin{bmatrix} 15 & -25 & -5 \\ 0 & 0 & -75 \\ 20 & 50 & 35 \end{bmatrix}$$

$$= \begin{bmatrix} 25 & 25 & 25 \\ 0 & 0 & -75 \\ 0 & 50 & 25 \end{bmatrix}.$$

To complete the reduction a rotation $R(2, 3)$ with $\cos\theta = 0$, $\sin\theta = 1$ gives

$$A_3 = \begin{bmatrix} 25 & 25 & 25 \\ 0 & 50 & 25 \\ 0 & 0 & 75 \end{bmatrix}.$$

Therefore we have the decomposition

$$\begin{bmatrix} 12 & -20 & 41 \\ 9 & -15 & -63 \\ 20 & 50 & 35 \end{bmatrix} = R^T(1,2)\,R^T(1,3)\,R^T(2,3)\,A_3.$$

Since the product of elementary unitary matrices is a unitary matrix it follows that we have constructed a unitary matrix Q^* such that

$$Q^*A = U$$

or alternatively

$$A = QU.$$

Exercises

17. Show that it is possible to construct unitary matrices Q_1, Q_2, Q_3 so that the following decompositions exist

$$A = U_1Q_1, \quad A = Q_2L_2, \quad A = L_3Q_3,$$

where L_2, L_3 are lower triangular matrices.

18. A matrix X is said to be of Type I if $x_{ij} = 0$ for $i+j<n+1$ and of Type II if $x_{ij} = 0$ for $i+j>n+1$. Show that it is possible to decompose a matrix into a product of a unitary matrix and a matrix of Type I or of Type II.

The same type of decomposition can be achieved by using elementary Hermitian matrices in place of plane rotations (an alternative derivation to that of Section 3.4 is given here).

Let

$$P^{(r)} = I - 2\mathbf{w}^{(r)}\{\mathbf{w}^{(r)}\}^*$$

where $\{\mathbf{w}^{(r)}\}^* \mathbf{w}^{(r)} = 1$, and where

$$\{\mathbf{w}^{(r)}\}^T = (0, 0, \ldots, 0, w_{r+1}^{(r)}, \ldots, w_n^{(r)}).$$

Premultiplication of a matrix A by $P^{(r)}$ affects only rows $(r+1)$ to n of the matrix A and moreover the columns are treated independently. First of all let us consider premultiplication of the vector $\mathbf{x}$ by $P^{(r)}$, giving

$$\mathbf{y} = P^{(r)}\mathbf{x} = \mathbf{x} - 2\mathbf{w}^{(r)}\{\mathbf{w}^{(r)}\}^* \mathbf{x}.$$

We see that if we choose certain of the components of $\mathbf{w}^{(r)}$ to be the same multiple of the corresponding components of $\mathbf{x}$, then it is possible that the vector $\mathbf{y}$ may contain zeros in these positions. In particular, if we assume

$$P^{(r)} = I - G\mathbf{h}^{(r)}\{\mathbf{h}^{(r)}\}^* \tag{2}$$

where G is a constant, and if we define

$$\begin{aligned} S^2 &= \sum_{j=r+1}^{n} |x_j|^2, \\ T &= \{|x_{r+1}|^2 S^2\}^{1/2} \\ G &= (S^2+T)^{-1} \\ \{\mathbf{h}^{(r)}\}^T &= (0, 0, \ldots, 0, x_{r+1}(1+S^2/T), x_{r+2}, \ldots, x_n), \end{aligned} \tag{3}$$

then by expansion it may be verified that

$$\mathbf{y} = P^{(r)}\mathbf{x}$$

has zeros in positions $r+2$ to n and has elements 1 to r equal to the corresponding elements of $\mathbf{x}$. If the matrix A is real then we find that

$$\{\mathbf{h}^{(r)}\}^T = (0, 0, \ldots, 0, x_{r+1} \pm S, x_{r+2}, \ldots, x_n)$$

and that the $(r+1)$th element of $\mathbf{y}$ is given by $\mp S$. (We will in general only be concerned with real matrices.) The choice of the sign of S in the above depends on the sign of x_{r+1}, and the strategy generally adopted is to choose the sign of S to be the same as the sign of x_{r+1}. If we now in turn apply elementary Hermitian matrices $P^{(0)}, P^{(1)}, \ldots, P^{(n-2)}$ to the matrix A, choosing to annihilate the elements below the diagonal in the first, ..., $(n-1)$th column respectively, it may be verified that

$$P^{(n-2)} \ldots P^{(0)}A = U,$$

where U is an upper triangular matrix.

Example *Use elementary Hermitian matrices to carry out the above decomposition of the matrix*

$$A_0 = \begin{bmatrix} 12 & -20 & 41 \\ 9 & -15 & -63 \\ 20 & 50 & 35 \end{bmatrix}.$$

We seek a vector $\mathbf{w}$ such that

$$A_1 = (I-2\mathbf{w}\mathbf{w}^*)\, A_0$$

has its first column of the form $(y_1, 0, 0)^T$.

Using the analysis of this section or the simpler vector approach of Section 3.4 we see that the choice $\mathbf{w} = 1/\sqrt{650}\,(-13, -9, -20)^T$ gives the first column of A_1 to be $(25, 0, 0)^T$. The arithmetic in this approach is (manually) more cumbersome and is omitted. The process is repeated to obtain the annihilation of the last element of the second column of A_1. The details are left as an exercise.

Since the product of elementary Hermitian matrices is a unitary matrix it follows that this approach gives a decomposition of A into the form

$$A = QU,$$

analogous to our decomposition using plane rotations.

Exercise

19. Write a computer program which will construct the decomposition $A = QU$ using (*a*) plane rotations, with a test for zero elements, (*b*) elementary Hermitian matrices.

3.7 Elementary similarity transformations

In the previous section, it was shown that elementary unitary matrices could be used to decompose a matrix into a product of a unitary matrix and a triangular matrix. (In what follows, we will invariably assume the triangular matrix is of upper triangular form.) In this section, we consider similarity transformations of the matrix A by means of elementary unitary matrices. Consider first of all the use of plane rotations. The transformed matrix

$$R(j, k)\, AR^*(j, k)$$

differs from the matrix A only in the elements which lie in rows j and k and columns j and k.

Example *Consider the transformation of a* 5×5 *matrix* A *by a rotation in plane* (2, 4).

If we write those elements which are modified only in the premultiplication by $R(2, 4)$ as 'a', those which are modified only in the postmultiplication by $R^*(2, 4)$ as 'b', those modified in both pre- and postmultiplication as 'c' and those unchanged as 'x' we see that

$$R(2, 4)\, AR^*(2, 4) = \begin{bmatrix} x & b & x & b & x \\ a & c & a & c & a \\ x & b & x & b & x \\ a & c & a & c & a \\ x & b & x & b & x \end{bmatrix}.$$

We may choose our parameters of the rotation in order to make any of the a's, b's or c's zero. The differing strategies give rise to the various algorithms. In general a similarity transformation by a plane rotation of an $n \times n$ matrix A only modifies $4n-4$ elements.

Exercise

20. Let A be an $n \times n$ symmetric matrix and $R(j, k)$ be a plane rotation where $\alpha = \beta = 0$. Obtain the equations defining the elements of $\hat{A}$ where

$$\hat{A} = R(j, k)\, AR^*(j, k).$$

An alternative to the use of plane rotations is given by the similarity transformation

$$P^{(r)}AP^{(r)}$$

where $P^{(r)}$ is an elementary Hermitian matrix. The effect of this transformation is to modify the elements in row $r+1$ to n, and columns $r+1$ to n of the matrix A. Thus only the elements in the leading $r \times r$ submatrix of the matrix A remain unchanged.

Example *Consider the transformation of a* 5×5 *matrix* A *by the elementary Hermitian matrix* $P^{(2)}$.

If we now write those elements which are only affected in the premultiplication of A by $P^{(2)}$ as 'a', those only affected in postmultiplication as 'b', those affected in both as 'c' and those unchanged as 'x' we find

$$P^{(2)}AP^{(2)} = \begin{bmatrix} x & x & b & b & b \\ x & x & b & b & b \\ a & a & c & c & c \\ a & a & c & c & c \\ a & a & c & c & c \end{bmatrix}.$$

Again we may choose our elementary Hermitian matrices to give us an algorithm which successively reduces the matrix A to simpler form. We start the discussion of these algorithms in Chapter 8.

4

Methods for the dominant eigenvalue

4.1 Introduction

In many applications, not all of the eigenvalues and eigenvectors of a matrix are needed. In particular, it is common for only the eigenvalue which is largest in modulus to be required, and in this chapter we consider some methods for obtaining this *dominant* eigenvalue, along with the corresponding eigenvector.

The methods which we discuss are usually referred to as iterative methods: an arbitrary first approximation to the eigenvector corresponding to the dominant eigenvalue is successively improved until some required precision is reached. Convergence to the dominant eigenvalue is simultaneously obtained. We note in passing that all methods for obtaining eigenvalues of an $n \times n$ matrix are essentially iterative. However, methods making use of the similarity transformations described in the previous chapter are commonly referred to as direct methods or transformation methods. The iterative methods of this chapter (and also those of Chapter 5) are most useful in the treatment of large sparse matrices when good estimates of the eigenvectors are available.

4.2 The power method

Let A be an $n \times n$ matrix with linear elementary divisors whose eigenvalues satisfy

$$|\lambda_1| = |\lambda_2| = \ldots = |\lambda_r| > |\lambda_{r+1}| \geqslant \ldots \geqslant |\lambda_n|. \tag{1}$$

The eigenvalues $\lambda_1, \lambda_2, \ldots, \lambda_r$ will be referred to as the *dominant eigenvalues*. By assumption, there exist n linearly independent eigenvectors $\mathbf{x}_1, \mathbf{x}_2, \ldots, \mathbf{x}_n$, and any arbitrary vector $\mathbf{z}_0$ can be expressed in the form

$$\mathbf{z}_0 = \sum_{i=1}^{n} \alpha_i \mathbf{x}_i, \tag{2}$$

where α_i are scalars, not all zero.

Let us define the iterative scheme

$$\mathbf{z}_k = A\mathbf{z}_{k-1}, \quad k = 1, 2, 3, \ldots \tag{3}$$

where $\mathbf{z}_0$ is arbitrary. Then

$$\begin{aligned}\mathbf{z}_k &= A\mathbf{z}_{k-1} = A^2\mathbf{z}_{k-2} = \ldots = A^k\mathbf{z}_0 \\ &= \sum_{i=1}^{n} \alpha_i \lambda_i^k \mathbf{x}_i,\end{aligned} \tag{4}$$

where we have used equation (2).

Now provided that $\alpha_1, \alpha_2, \ldots, \alpha_r$ are not all zero, the right-hand side of equation (4) is ultimately dominated by the terms $\sum_{i=1}^{r} \alpha_i \lambda_i^k \mathbf{x}_i$. In particular, if $r = 1$, and we assume that $\alpha_1 \neq 0$, we have

$$\begin{aligned}\mathbf{z}_k &= \lambda_1^k \left\{\alpha_1 \mathbf{x}_1 + \sum_{i=2}^{n} \alpha_i (\lambda_i/\lambda_1)^k \mathbf{x}_i\right\} \\ &= \lambda_1^k \{\alpha_1 \mathbf{x}_1 + \boldsymbol{\epsilon}_k\}\end{aligned} \tag{5}$$

for k sufficiently large, where $\boldsymbol{\epsilon}_k$ is a vector with very small elements. The vector $\mathbf{z}_k$ is an approximation to the unnormalized eigenvector $\mathbf{x}_1$, and is accurate if $\|\boldsymbol{\epsilon}_k\|$ is sufficiently small. This observation is the basis for the simple power method for computing the dominant eigenvalue.

Since

$$\mathbf{z}_{k+1} = \lambda_1^{k+1}\{\alpha_1 \mathbf{x}_1 + \boldsymbol{\epsilon}_{k+1}\},$$

it follows that for any i

$$\begin{aligned}\frac{(\mathbf{z}_{k+1})_i}{(\mathbf{z}_k)_i} &= \lambda_1 \left\{\frac{\alpha_1(\mathbf{x}_1)_i + (\boldsymbol{\epsilon}_{k+1})_i}{\alpha_1(\mathbf{x}_1)_i + (\boldsymbol{\epsilon}_k)_i}\right\} \\ &\rightarrow \lambda_1 \quad \text{as} \quad k \rightarrow \infty.\end{aligned} \tag{6}$$

(To avoid a double subscript, $(\mathbf{z}_k)_i$ is used to denote the ith component of $\mathbf{z}_k$.)

The rate of convergence will depend on the constants α_i, but more essentially on the ratios $|\lambda_2/\lambda_1|, |\lambda_3/\lambda_1|, \ldots, |\lambda_n/\lambda_1|$, and the smaller these ratios the faster will be the convergence. In particular, if $|\lambda_2/\lambda_1|$ is close to unity, then convergence is likely to be very slow.

In practice, the elements of $\mathbf{z}_k$ are scaled at each step, and equation (3) is replaced by the pair of equations

$$\begin{aligned}\mathbf{y}_k &= A\mathbf{z}_{k-1} \\ \mathbf{z}_k &= \mathbf{y}_k/\gamma_k\end{aligned}$$

where γ_k is the component of $\mathbf{y}_k$ of greatest modulus. In this case

$$\mathbf{z}_k \rightarrow \mathbf{x}_1/\|\mathbf{x}_1\|_\infty$$

and

$$\gamma_k \rightarrow \lambda_1 \text{ as } k \rightarrow \infty.$$

Example *Find the dominant eigenvalue of the matrix*

$$A = \begin{bmatrix} 2 & 3 & 2 \\ 10 & 3 & 4 \\ 3 & 6 & 1 \end{bmatrix}.$$

Taking $\mathbf{z}_0 = [0 \quad 0 \quad 1]^T$, we have

k		$\mathbf{z}_k^T$		$\|\mathbf{y}_k\|_\infty$
0	0	0	1	1
1	0·5	1·0	0·25	4
2	0·5	1·0	0·8611	9
3	0·5	1·0	0·7306	11·44
4	0·5	1·0	0·7535	10·9224
5	0·5	1·0	0·7493	11·0140
6	0·5	1·0	0·7501	10·9972
7	0·5	1·0	0·7500	11·0004
8	0·5	1·0	0·7500	11·0000

Since the exact eigenvalues are 11, -3 and -2, the ratios $|\lambda_2/\lambda_1|$ and $|\lambda_3/\lambda_1|$ are small, accounting for the fast convergence of the method in this case.

Now suppose that $r > 1$, and that equation (1) is satisfied with

$$\lambda_1 = \lambda_2 = \ldots = \lambda_r.$$

Then we have

$$\mathbf{z}_k = \lambda_1^k \left\{ \sum_{i=1}^{r} \alpha_i \mathbf{x}_i + \sum_{i=r+1}^{n} \alpha_i (\lambda_i/\lambda_1)^k \, \mathbf{x}_i \right\}$$

$$= \lambda_1^k \left\{ \sum_{i=1}^{r} \alpha_i \mathbf{x}_i + \boldsymbol{\epsilon}_k \right\}, \tag{7}$$

for sufficiently large k, where $\boldsymbol{\epsilon}_k$ is again a vector with very small elements. Thus convergence of the power method is not affected, and the iterates $\mathbf{z}_k$ tend to a vector which is some linear combination of the eigenvectors corresponding to λ_1. Thus the power method will only supply *one* eigenvector corresponding to a multiple dominant eigenvalue, for each $\mathbf{z}_0$.

The iterative procedure breaks down, however, if there are a number of unequal eigenvalues of the same modulus. This breakdown is characterized by the failure of the iterates to converge, and by changes in sign of the approximations to λ_1. We consider the case where we have a pair of complex conjugate dominant eigenvalues λ_1 and $\lambda_2 = \bar{\lambda}_1$. In this case, the effect of $\bar{\lambda}_1$ and its corresponding eigenvector $\bar{\mathbf{x}}_1$ will not be damped out as the iteration proceeds, and instead of equation (5), we will have

$$\mathbf{z}_k = \alpha_1 \lambda_1^k \mathbf{x}_1 + \bar{\alpha}_1 \bar{\lambda}_1^k \bar{\mathbf{x}}_1 + \boldsymbol{\epsilon}_k.$$

Let λ_1 and $\bar{\lambda}_1$ be the roots of the equation

$$\lambda^2+b\lambda+c = 0, \quad b, c, \text{ real}. \tag{8}$$

Then provided that the quantities ϵ_{k+i}, $i = 0, 1, 2$, are small enough to be neglected, we will have

$$\mathbf{z}_{k+2}+b\mathbf{z}_{k+1}+c\mathbf{z}_k = 0, \tag{9}$$

and the constants b and c can be found from any two equations of the set. A better procedure is to use all the n equations of (9) and calculate b and c such that

$$\sum_{i=1}^{n} [(\mathbf{z}_{k+2})_i+b(\mathbf{z}_{k+1})_i+c(\mathbf{z}_k)_i]^2$$

is a minimum. This is a linear least squares problem. The corresponding eigenvectors can then be found from any two successive vectors $\mathbf{z}_k$ and $\mathbf{z}_{k+1}$.

This process can clearly be generalized to deal with any number of unequal dominating eigenvalues of the same modulus, or indeed eigenvalues $\lambda_1, \lambda_2, \ldots, \lambda_m$ (real or complex) which satisfy

$$|\lambda_1| \geqslant |\lambda_2| \geqslant \ldots \geqslant |\lambda_m| \gg |\lambda_{m+1}| \geqslant \ldots \geqslant |\lambda_n|.$$

This procedure appears particularly promising when $|\lambda_1|, \ldots, |\lambda_m|$ are close. However, the resulting polynomials of degree m corresponding to equation (8) can be very ill-conditioned (in the sense that small changes in the coefficients can give arbitrarily large changes in the roots), and better results can be obtained using the methods described in Section 5.3 of the next chapter.

Exercises

1. Taking $\mathbf{z}_0 = [1, 1, 1]^T$, apply the power method to find to four decimal places the dominant eigenvalues and eigenvectors of the matrices

(*a*)
$$\begin{bmatrix} 6 & 2 & 1 \\ 2 & 3 & 1 \\ 1 & 1 & 1 \end{bmatrix}$$

(*b*)
$$\begin{bmatrix} 0 & 1 & 1 \\ -1 & 0 & 1 \\ -1 & -1 & 0 \end{bmatrix}$$

2. When A has non-linear elementary divisors, the simple power method can still converge, although extremely slowly. Illustrate this for the matrix

$$A = \begin{bmatrix} 1 & 1 \\ 0 & 1 \end{bmatrix},$$

taking $\mathbf{z}_0 = [1, 1]^T$.

3. The matrix

$$A = \begin{bmatrix} 2 & 1 & -1 \\ 1 & 2 & -1 \\ 1 & -1 & 2 \end{bmatrix}$$

has eigenvalues 3, 2 and 1. Explain the behaviour of the power method when applied using the initial approximation $\mathbf{z}_0 = (1, 0, 0)^T$. Repeat the calculation taking $\mathbf{z}_0 = (1, 0, \epsilon)^T$, where ϵ is a small number.

4.3 Shift of origin

Most practical techniques based on the power method permit the use of a device for the acceleration of convergence. The simplest of these is based on the observation that for a single dominant eigenvalue, the rate of convergence depends essentially on $|\lambda_2/\lambda_1|$. If we consider the application of the power method to the matrix $A-pI$ instead of A, then since the eigenvalues of this new matrix are $\lambda_i - p$, provided that $\lambda_1 - p$ still dominates, the rate of convergence is governed by $|(\lambda_2-p)/(\lambda_1-p)|$. For suitable p, this ratio can be made much smaller than $|\lambda_2/\lambda_1|$.

This device is commonly referred to as *shift of origin*, and for certain distributions of the eigenvalues can be quite effective. For example, for a matrix of order four with eigenvalues $\lambda_j = 15-j$, the usual iteration gives convergence to $\mathbf{x}_1$ at a rate governed essentially by $(13/14)^k$. However if we choose $p = 12$ and treat the matrix $A-pI$, then the convergence rate is governed by $(\frac{1}{2})^k$.

Although the value of p can often be advantageously chosen, it is difficult to design an automatic procedure for doing this. It is possible to successfully use *ad hoc* techniques however, particularly if the eigenvalues are all real. We illustrate by an example.

Example *Consider the application of the power method to the matrix*

$$A = \begin{bmatrix} -3 & 1 & 0 \\ 1 & -3 & -3 \\ 0 & -3 & 4 \end{bmatrix}$$

with initial approximation $\mathbf{z}_0 = [0 \quad 0 \quad 1]^T$.

The first six iterates are

k		$\mathbf{z}_k^T$		$\|\mathbf{y}_k\|_\infty$
0	0	0	1	1
1	0	−0·75	1	4
2	−0·12	−0·12	1	6·25
3	0·0550	−0·6330	1	4·36
4	−0·0135	−0·1773	1	5·899
5	−0·0302	−0·5476	1	4·5319
6	−0·0801	−0·2459	1	5·6428

Convergence is clearly very slow, and further it is oscillatory. If we assume that the rate of convergence is governed only by $(\lambda_2/\lambda_1)^k$, then we predict an eigenvalue λ_2 close to λ_1 in magnitude, but opposite in sign. Let us take $p = -4$ (thereby shifting λ_1 to about 10 and λ_2 to about -2) and apply the power method to the matrix

$$A+4I = \begin{bmatrix} 1 & 1 & 0 \\ 1 & 1 & -3 \\ 0 & -3 & 8 \end{bmatrix}.$$

In this case, the iteration proceeds as follows:

k	$\mathbf{z}_k^T$			$\|\mathbf{y}_k\|_\infty$
0	0	0	1	1
1	0	$-0{\cdot}375$	$1{\cdot}0$	8
2	$-0{\cdot}0411$	$-0{\cdot}3699$	$1{\cdot}0$	$9{\cdot}125$
3	$-0{\cdot}0451$	$-0{\cdot}3744$	$1{\cdot}0$	$9{\cdot}1097$
4	$-0{\cdot}0460$	$-0{\cdot}3748$	$1{\cdot}0$	$9{\cdot}1232$
5	$-0{\cdot}0461$	$-0{\cdot}3749$	$1{\cdot}0$	$9{\cdot}1244$
6	$-0{\cdot}0461$	$-0{\cdot}3749$	$1{\cdot}0$	$9{\cdot}1247$

Thus, after six iterations, we have $\lambda_1 = 5{\cdot}125$ correct to three decimal places.

Exercises

4. Show that the value of p which will give the fastest convergence to $\mathbf{x}_1$ for the general distribution of real eigenvalues

$$\lambda_1 > \lambda_2 \geqslant \lambda_3 \geqslant \ldots \geqslant \lambda_n$$

is $p = \frac{1}{2}(\lambda_2 + \lambda_n)$.

5. Calculate to four places of decimals the dominant eigenvalue of the matrix

$$A = \begin{bmatrix} -11 & 11 & 1 \\ 11 & 9 & -2 \\ 1 & -2 & 13 \end{bmatrix}$$

by directly applying the power method, and predicting a suitable shift of origin.

4.4 Aitken's acceleration device

Another method which can be used to speed up the convergence of the simple power method was described in 1937 by A. C. Aitken. The device is applicable to any process which is converging linearly (i.e. as in equation (10) below), and can best be described as follows (see also Morris (1974)).

Let x_i, $i = 0, 1, 2, \ldots$, be a sequence of approximations to α such that

$$x_i = \alpha + \epsilon_i, \quad i = 0, 1, 2, \ldots$$

where

$$\epsilon_i = K_i \epsilon_{i-1}, \quad i = 1, 2, \ldots \tag{10}$$

and the K_i are constants such that $|K_i| < 1$.

Now $K_i \to K$ (the asymptotic error constant) as $i \to \infty$ and so for sufficiently large k, we have

$$\frac{x_{k+2} - \alpha}{x_{k+1} - \alpha} \approx \frac{x_{k+1} - \alpha}{x_k - \alpha},$$

and so

$$\alpha \approx \frac{x_k x_{k+2} - x_{k+1}^2}{x_k - 2x_{k+1} + x_{k+2}}. \tag{11}$$

Now let $\mathbf{y}_k$, $\mathbf{y}_{k+1}$ and $\mathbf{y}_{k+2}$ be three successive iterates obtained by the power method, and suppose that a stage has been reached where essentially

$$\mathbf{y}_k = \mathbf{x}_1 + \epsilon \mathbf{x}_2$$

where ϵ is small.

Now if $\mathbf{x}_1$ and $\mathbf{x}_2$ are normalized so that their largest elements are unity, then for sufficiently small ϵ, the largest element of $\mathbf{y}_k$ will be in the same position as the largest element of $\mathbf{x}_1$. If the element of $\mathbf{x}_2$ in this position is d, where $|d| \leqslant 1$, then we have, on normalizing $\mathbf{y}_k$, $\mathbf{y}_{k+1}$ and $\mathbf{y}_{k+2}$,

$$\mathbf{z}_k = \frac{\mathbf{x}_1 + \epsilon \mathbf{x}_2}{1 + \epsilon d}, \tag{12}$$

$$\mathbf{z}_{k+1} = \frac{\lambda_1 \mathbf{x}_1 + \epsilon \lambda_2 \mathbf{x}_2}{\lambda_1 + \epsilon d \lambda_2}, \tag{13}$$

$$\mathbf{z}_{k+2} = \frac{\lambda_1^2 \mathbf{x}_1 + \epsilon \lambda_2^2 \mathbf{x}_2}{\lambda_1^2 + \epsilon d \lambda_2^2}. \tag{14}$$

The Aitken acceleration process can then be applied to these three iterates to produce an approximation $\mathbf{w}$ such that

$$\mathbf{w}_i = \frac{(\mathbf{z}_k)_i (\mathbf{z}_{k+2})_i - (\mathbf{z}_{k+1})_i^2}{(\mathbf{z}_k)_i - 2(\mathbf{z}_{k+1})_i + (\mathbf{z}_{k+2})_i}, \quad i = 1, 2, \ldots, n.$$

Substituting from equations (12), (13) and (14) it follows that

$$\mathbf{w} = \mathbf{x}_1 + 0(\epsilon^2), \tag{15}$$

(where the notation $c = 0(\epsilon^2)$ means that $\lim_{\varepsilon \to 0} c/\epsilon^2$ remains finite).

This procedure is not seriously affected by rounding errors, but is rather difficult to automate on a computer.

An analogous process can be applied to improve the current estimate of the eigenvalue. We assume that λ_1 and λ_2 are real. Then using equation (6) (absorbing the coefficients α_i into the vectors $\mathbf{x}_i$) it may be shown that

$$\frac{(\mathbf{z}_{k+1})_i}{(\mathbf{z}_k)_i} = \lambda_1 + \left(\frac{\lambda_2}{\lambda_1}\right)^k (\lambda_2 - \lambda_1)\frac{(\mathbf{x}_2)_i}{(\mathbf{x}_1)_i} + \epsilon_i.$$

Writing $(\mathbf{z}_{k+1})_i/(\mathbf{z}_k)_i = \mu_k$, the current estimate of λ_1, and assuming ϵ_i to be negligible, we have

$$\mu_k - \lambda_1 = \left(\frac{\lambda_2}{\lambda_1}\right)^k (\lambda_2 - \lambda_1)\frac{(\mathbf{x}_2)_i}{(\mathbf{x}_1)_i}, \tag{16}$$

and the usual Aitken formula (11) can be applied to improve the estimates μ_k, μ_{k+1} and μ_{k+2}.

Exercises

6. Derive equation (15).

7. Apply the Aitken acceleration process to speed the convergence of the iterates $\mathbf{z}_k$ of exercise 5 of Section 4.3.

4.5 The Rayleigh quotient

Let A be an $n \times n$ real symmetric matrix. Then for any non-trivial vector $\mathbf{x}$ the quantity

$$\frac{\mathbf{x}^* A \mathbf{x}}{\mathbf{x}^* \mathbf{x}}$$

is called the *Rayleigh quotient* corresponding to $\mathbf{x}$. Further, in Theorem 2.5, we showed that if λ_1 is the largest eigenvalue of A, then

$$\lambda_1 = \max_{\mathbf{x} \neq 0} \frac{\mathbf{x}^* A \mathbf{x}}{\mathbf{x}^* \mathbf{x}}, \tag{17}$$

and the maximum is attained when $\mathbf{x}$ is the eigenvector corresponding to λ_1. In this case, the computation of λ_1 is an optimization problem and, for example, gradient methods are available for its solution. The problem of determining λ_1 from equation (17) has been considered in detail by Ostrowski.

Our main interest in the Rayleigh quotient is in its use for the acceleration of convergence of the dominant eigenvalue given by the power method applied to a real symmetric matrix. In this case, the eigenvectors can be chosen to satisfy

$$\mathbf{x}_i^T \mathbf{x}_j = \delta_{ij}, \tag{18}$$

i.e. they are orthonormal.

Now consider the application of the power method in its simplest form as in equation (3). Then

$$\mathbf{z}_k^T A \mathbf{z}_k = \mathbf{z}_k^T \mathbf{z}_{k+1}$$

$$= \sum_{i=1}^{n} \alpha_i^2 \lambda_i^{2k+1}$$

and

$$\mathbf{z}_k^T \mathbf{z}_k = \sum_{i=1}^{n} \alpha_i^2 \lambda_i^{2k},$$

where we have used equations (4) and (18).

Thus

$$\frac{\mathbf{z}_k^T A \mathbf{z}_k}{\mathbf{z}_k^T \mathbf{z}_k} = \lambda_1 + 0\left(\left(\frac{\lambda_2}{\lambda_1}\right)^{2k}\right), \tag{19}$$

and so by comparison with equation (16), the Rayleigh quotient corresponding to $\mathbf{z}_k$ will generally give a better approximation to λ_1 than the power method itself.

Example Calculate some Rayleigh quotients from the sequence of iterates $\mathbf{z}_k$ *of the shifted worked example of Section* 4.3.

We have

$$\frac{\mathbf{z}_3^T A \mathbf{z}_3}{\mathbf{z}_3^T \mathbf{z}_3} = 9{\cdot}1247$$

$$\frac{\mathbf{z}_4^T A \mathbf{z}_4}{\mathbf{z}_4^T \mathbf{z}_4} = 9{\cdot}1246.$$

These values are clearly better than the normal power method approximations.

Exercises

8. (*a*) Show that the Rayleigh quotient gives a better approximation to λ_1 than the power method by evaluation for some of the iterates obtained in exercise 5 of Section 4.3.

(*b*) Compare the results with those obtained by applying the Aitken acceleration process for the eigenvalues.

9. The matrix

$$\begin{bmatrix} -1 & 2 & 1 \\ 2 & -4 & 1 \\ 1 & 1 & -6 \end{bmatrix}$$

has eigenvalues near 0 and -5. Apply the power method to find the dominant eigenvalue correct to two places of decimals. Accelerate the convergence of the iteration using one of the devices described.

5

Methods for the subdominant eigenvalues

5.1 Introduction

In the previous chapter, procedures based on the power method were presented for finding the dominant eigenvalue of a matrix. In addition, an indication was given how eigenvalues $\lambda_1, \lambda_2, \ldots, \lambda_m$ satisfying

$$|\lambda_1| \geqslant |\lambda_2| \geqslant \ldots \geqslant |\lambda_m| \gg |\lambda_{m+1}| \geqslant \ldots \geqslant |\lambda_n|$$

can be obtained by an extension of the technique for dealing with a pair of dominant complex conjugate eigenvalues. However the procedure involves finding the roots of ill-conditioned polynomials, and is not recommended.

In this chapter, we consider some stable iterative methods for finding the eigenvalues $\lambda_1, \lambda_2, \ldots, \lambda_m$. As mentioned previously, iterative methods are attractive in particular if the matrix is large and sparse, with good estimates of the eigenvectors available.

5.2 Deflation

We will assume that the values $|\lambda_1|, |\lambda_2|, \ldots, |\lambda_m|$ are well separated. Then λ_1 can be obtained by the methods of the previous chapter, and it remains to compute the *subdominant* eigenvalues $\lambda_2, \ldots, \lambda_m$.

Some of the most useful procedures for finding the eigenvalues in this case are based on the derivation of a new matrix from the original matrix by some form of 'deflation' process. The new matrix is constructed in such a way that essentially it contains only the remaining unknown eigenvalues of the original matrix. Repeated application of this deflation process enables the subdominant eigenvalues and corresponding eigenvectors to be computed sequentially. The most popular deflation processes are those based on similarity transformations. To describe the procedure, we begin by assuming that the dominant eigenvalue λ_1 and corresponding eigenvector $\mathbf{x}_1$ of the matrix A_1 have been calculated. Now let R_1 be a non-singular matrix such that

$$R_1\mathbf{x}_1 = k\mathbf{e}_1 \tag{1}$$

where $k \neq 0$. We defer for the moment consideration of how R_1 is obtained. Then

$$R_1A_1(R_1^{-1}R_1)\,\mathbf{x}_1 = \lambda_1 R_1\mathbf{x}_1,$$

and thus

$$R_1A_1R_1^{-1}\mathbf{e}_1 = \lambda_1\mathbf{e}_1.$$

Hence we may write

$$A_2 = R_1A_1R_1^{-1} = \begin{bmatrix} \lambda_1 & \boldsymbol{\gamma}^T \\ 0 & B_2 \end{bmatrix}$$

where B_2 is a matrix of order $n-1$ and $\boldsymbol{\gamma}$ is a vector of $n-1$ elements. Now since A_2 has the same eigenvalues as A_1, B_2 possesses eigenvalues $\lambda_2, \lambda_3, \ldots, \lambda_n$. Thus we can work with B_2 to obtain the next eigenvalue λ_2 and corresponding eigenvector $\mathbf{y}_2$ satisfying

$$B_2\mathbf{y}_2 = \lambda_2\mathbf{y}_2. \tag{2}$$

It remains to find the eigenvector $\mathbf{x}_2$ of A_1 corresponding to λ_2. Let $\mathbf{z}_2$ be the eigenvector of A_2 corresponding to λ_2. Then

$$\begin{bmatrix} \lambda_1 & \boldsymbol{\gamma}^T \\ 0 & B_2 \end{bmatrix} \mathbf{z}_2 = \lambda_2\mathbf{z}_2, \tag{3}$$

and using equation (2) we can take

$$\mathbf{z}_2 = \begin{bmatrix} \alpha \\ \mathbf{y}_2 \end{bmatrix} \tag{4}$$

where α is a scalar given by

$$(\lambda_1 - \lambda_2)\,\alpha + \boldsymbol{\gamma}^T\mathbf{y}_2 = 0.$$

Finally

$$\mathbf{x}_2 = R_1^{-1}\mathbf{z}_2$$

and we have the required eigenvector.

Continuing in this way, we may obtain the remaining subdominant eigenvalues and corresponding eigenvectors of A_1. We note that successive deflations will give, in the limit, a similarity reduction of A to upper triangular form.

We now consider two possible ways of choosing the matrix R_1 which lead to deflation processes which are numerically stable. The simplest of these involves choosing R_1 to be the product of an elementary lower triangular matrix L_1 and a permutation matrix $I_{1,p}$, where p is such that $(\mathbf{x}_1)_p$ is the largest element in modulus of $\mathbf{x}_1$. Thus we have

$$\mathbf{y} = I_{1,p}\mathbf{x}_1$$

$$L_1\mathbf{y} = k\mathbf{e}_1$$

$$\text{where } L_1 = \begin{bmatrix} 1 & & & \\ -y_2/y_1 & 1 & & \\ \vdots & & \ddots & \\ -y_n/y_1 & & & 1 \end{bmatrix}$$

and $k = y_1 = (\mathbf{x}_1)_p$.

The introduction of the permutation matrix $I_{1,p}$ effectively defines a pivoting process which enables the elements of R_1 to be bounded in modulus by unity, and ensures stability of the resulting algorithm by an argument analogous to that used for Gaussian elimination.

Example *Illustrate the steps involved by considering the deflation of the matrix*

$$A_1 = \begin{bmatrix} 2 & 3 & 2 \\ 10 & 3 & 4 \\ 3 & 6 & 1 \end{bmatrix}$$

assuming a dominant eigenvalue $\lambda_1 = 11{\cdot}0$ *with corresponding eigenvector*

$$\mathbf{x}_1 = (0{\cdot}5, 1{\cdot}0, 0{\cdot}75)^T.$$

Here we have $p = 2$ and so

$$L_1 = \begin{bmatrix} 1 & & \\ -0{\cdot}5 & 1 & \\ -0{\cdot}75 & 0 & 1 \end{bmatrix}.$$

Thus

$$A_2 = L_1 I_{1,2} A_1 I_{1,2} L_1^{-1}$$

$$= \begin{bmatrix} 1 & & \\ -0{\cdot}5 & 1 & \\ -0{\cdot}75 & 0 & 1 \end{bmatrix} \begin{bmatrix} 3 & 10 & 4 \\ 3 & 2 & 2 \\ 6 & 3 & 1 \end{bmatrix} \begin{bmatrix} 1 & & \\ 0{\cdot}5 & 1 & \\ 0{\cdot}75 & 0 & 1 \end{bmatrix}$$

$$= \begin{bmatrix} 3 & 10 & 4 \\ 1{\cdot}5 & -3 & 0 \\ 3{\cdot}75 & -4{\cdot}5 & -2 \end{bmatrix} \begin{bmatrix} 1 & & \\ 0{\cdot}5 & 1 & \\ 0{\cdot}75 & 0 & 1 \end{bmatrix}$$

$$= \begin{bmatrix} 11 & 10 & 4 \\ 0 & -3 & 0 \\ 0 & -4{\cdot}5 & -2 \end{bmatrix}.$$

(Note that the postmultiplication by L_1^{-1} need not actually be performed, as we know the first column of A_2 to be $\lambda_1\mathbf{e}_1$.)

Hence

$$B_2 = \begin{bmatrix} -3 & 0 \\ -4\cdot 5 & -2 \end{bmatrix}$$

and it follows that the remaining eigenvalues are -3 and -2.

The second procedure takes R_1 to be a unitary matrix (orthogonal if A_1 is real), for example the elementary Hermitian matrix $I-2\mathbf{w}\mathbf{w}^*$ introduced in Section 3.4. An explicit expression for $\mathbf{w}$ such that equation (1) is satisfied in this case can be obtained from expressions in that section; for example when A_1 is real we can take

$$\mathbf{w} = \frac{\mathbf{x}_1 - k\mathbf{e}_1}{\sqrt{2k(k-\mathbf{e}_1^T\mathbf{x}_1)}},$$

$$k^2 = \mathbf{x}_1^T\mathbf{x}_1.$$

Alternatively, we may choose R_1 to be the product of rotations in the planes $(1, 2), (1, 3), \ldots, (1, n)$ (see Section 3.3). However, this latter process involves twice as many multiplications as that using elementary Hermitian matrices.

The unitary deflation processes are extremely stable, although the non-unitary process described above requires fewer multiplications. However this may be but a small fraction of the total work involved in finding successive eigenvalues and eigenvectors, and in this case, the use of unitary transformations is to be recommended.

Exercises

1. Show that approximately n^2 multiplications are involved in the stabilized non-unitary deflation process, and approximately $4n^2$ and $8n^2$ multiplications respectively are involved in the use of elementary Hermitian matrices, and plane rotation matrices.

2. Use the results of this section to prove Theorem 2.2.

3. Use one of the deflation processes described in this section to compute all the eigenvalues and eigenvectors of the matrix in exercise 5 of Section 4.3, correct to three decimal places.

4. Assume that we know eigenvalues λ_1, λ_2 and corresponding eigenvectors $\mathbf{x}_1$, $\mathbf{x}_2$ of a matrix A. Define a suitable deflation process which gives in one step a matrix with the remaining eigenvalues of A. (Hint: Let X be a $n\times 2$ matrix with columns $\mathbf{x}_1$ and $\mathbf{x}_2$, and choose R_1 such that

$$R_1X = \begin{bmatrix} T \\ 0 \end{bmatrix}$$

where T is 2×2 upper triangular.)

5.3 Simultaneous iteration for real symmetric matrices

The deflation processes of Section 5.2 require that the eigenvalues (and eigenvectors) be computed one at a time, and therefore they will break down if at some stage we have multiple dominant eigenvalues or a dominant complex conjugate pair. Suitable redefinition of the deflation process could enable more than one eigenvalue to be eliminated at each step (see, for example, exercise 4, Section 5.2). However it is possible to adopt a more general iterative approach to the computation of the subdominant eigenvalues, which is also useful if these eigenvalues are not well separated in modulus.

The procedure is based on iterating simultaneously with a number of trial vectors. The original method, called 'Bi-iteration', is due to F. L. Bauer, and this method has subsequently been specialized to symmetric positive definite matrices by Rutishauser (1969), and to symmetric matrices by Clint and Jennings (1970). The basic difference in the methods is the way in which the set of trial vectors is updated at each iteration. We now give an account of the method of Clint and Jennings. The reader is referred to their paper for additional details.

Let A be a real symmetric matrix of order n whose eigenvalues satisfy

$$|\lambda_1| \geqslant |\lambda_2| \geqslant \ldots \geqslant |\lambda_n|,$$

and let the corresponding eigenvectors be such that

$$\mathbf{x}_i^T \mathbf{x}_j = \delta_{ij}.$$

Let the $n \times m$ matrix

$$U = [\mathbf{u}_1, \mathbf{u}_2, \ldots, \mathbf{u}_m]$$

have as columns a set of m trial eigenvectors approximating $\mathbf{x}_1, \mathbf{x}_2, \ldots, \mathbf{x}_m$, where in general $1 < m \ll n$, and let the vectors be normalized so that

$$U^T U = I. \tag{5}$$

Then we define the $n \times m$ matrix

$$V = [\mathbf{v}_1, \mathbf{v}_2, \ldots, \mathbf{v}_m]$$

by

$$V = AU. \tag{6}$$

If the trial vectors are exact eigenvectors of A, then the matrix

$$B = U^T V = U^T A U \tag{7}$$

will be diagonal with m eigenvalues of A as elements. In general, the size of the off-diagonal elements will depend on the degree of interaction between the vectors $\mathbf{u}_i$ and $\mathbf{v}_j$, i.e. how near $\mathbf{u}_i^T \mathbf{v}_j$ is to zero for $i \neq j$, and for this reason B is referred to as the interaction matrix.

Let $X = [\mathbf{x}_1, \mathbf{x}_2, \ldots, \mathbf{x}_n]$. Then we can write

$$U = XC, \tag{8}$$

where C is an $n \times m$ matrix of coefficients which must satisfy

$$C^T C = I. \tag{9}$$

Thus

$$B = C^T \Lambda C, \tag{10}$$

where

$$\Lambda = \text{diag}\,(\lambda_1, \lambda_2, \ldots, \lambda_n).$$

If $m = n$, then it follows that C is the matrix of eigenvectors of B. Further, by equation (7) B is an orthogonal transformation of A which will be nearly diagonal if the trial vectors are fairly accurate approximations to the eigenvectors of A, and can thus be readily reduced to diagonal form by the method of Jacobi (see Chapter 7).

When $m < n$, however, the matrix C is not square, and a less direct approach is adopted. Write

$$U = X_a C_a + X_b C_b, \tag{11}$$

where

$$X_a = [\mathbf{x}_1, \mathbf{x}_2, \ldots, \mathbf{x}_m],$$

and

$$X_b = [\mathbf{x}_{m+1}, \ldots, \mathbf{x}_n],$$

and C_a and C_b are coefficient matrices, with C_a square. Then

$$B = C_a^T \Lambda_a C_a + C_b^T \Lambda_b C_b, \tag{12}$$

where

$$\Lambda_a = \text{diag}\,(\lambda_1, \lambda_2, \ldots, \lambda_m),$$

$$\Lambda_b = \text{diag}\,(\lambda_{m+1}, \ldots, \lambda_n).$$

Then equation (9) becomes

$$C_a^T C_a + C_b^T C_b = I. \tag{13}$$

If the coefficients C_b are assumed small compared with C_a, then the eigenvalues of B are seen to be approximately the m largest eigenvalues of A, and the matrix of eigenvectors of B (say P) is approximately C_a^T. This observation gives the basis for an iterative procedure.

Suppose that we express V in the form

$$V = X_a \Lambda_a C_a + X_b \Lambda_b C_b. \tag{14}$$

Then, ignoring the contributions involving C_b in equations (13) and (14), a closer approximation to the first m eigenvectors is given by

$$W = VP$$

since

$$VP \approx VC_a^T \approx X_a \Lambda_a.$$

In general, W will not be an orthogonal matrix, and so the new set of trial vectors $\hat{\mathbf{U}}$ must be found by an orthogonalization process. The technique normally used is called the *Gram–Schmidt orthogonalization process*, which may be applied to any set of linearly independent vectors. Columns $\hat{\mathbf{u}}_j$ *of* $\hat{\mathbf{U}}$ are constructed iteratively, with $\hat{\mathbf{u}}_j$ obtained as a linear combination of $\mathbf{w}_j$ (the jth column of W) and $\hat{\mathbf{u}}_i$, $i = 1, 2, \ldots, j-1$. The multipliers are chosen so that

$$\hat{\mathbf{u}}_j^T \hat{\mathbf{u}}_i = \delta_{ij}, \quad i = 1, 2, \ldots, j.$$

The full iteration procedure can be summarized as follows:

(i) Compute $V = AU$.
(ii) Compute $B = U^T V$.
(iii) Calculate the eigenvalues and the matrix P of eigenvectors of B.
(iv) Compute $W = VP$.
(v) Form U by an orthogonalization process applied to W.
(vi) Test for convergence by examining the discrepancy between successive trial vectors, and return to step (i) if necessary.

As in the power method, step (i) has the effect of increasing the dominance of the eigenvectors corresponding to $\lambda_1, \lambda_2, \ldots, \lambda_m$, and hence at each iteration the matrix C_b of coefficients will be reduced. The eigenvalues of B thus tend to the m largest eigenvalues of A, and the columns of the matrix U converge to the corresponding eigenvectors of A.

With reference to step (iii), we note that if $m \ll n$, then the evaluation of the eigensystem of B will be numerically much shorter than that of A, and any numerical procedure may be used. Further, since the results are to be fed back into an iterative cycle, high accuracy is not essential initially. In the latter stages of the iteration, B will be near diagonal, and the Jacobi method described in Chapter 7 is to be recommended.

Example *Using the method described above, find* λ_1, λ_2 *and the corresponding eigenvectors of the matrix*

$$A = \begin{bmatrix} 4 & 1 & 4 \\ 1 & 10 & 1 \\ 4 & 1 & 10 \end{bmatrix}.$$

Taking initially

$$U = \begin{bmatrix} 1 & 0 \\ 0 & 1 \\ 0 & 0 \end{bmatrix},$$

we first calculate

$$V = \begin{bmatrix} 4 & 1 \\ 1 & 10 \\ 4 & 1 \end{bmatrix}$$

followed by

$$B = \begin{bmatrix} 4\cdot 0 & 1\cdot 0 \\ 1\cdot 0 & 10\cdot 0 \end{bmatrix}.$$

This matrix has eigenvalues $\mu_1 = 10{\cdot}16$ and $\mu_2 = 3{\cdot}84$ with matrix of eigenvectors

$$P = \begin{bmatrix} 0\cdot 16 & 0\cdot 99 \\ 0\cdot 99 & -0\cdot 16 \end{bmatrix}$$

where $P^T P = I$.

Thus

$$W = \begin{bmatrix} 1\cdot 63 & 3\cdot 79 \\ 10\cdot 03 & -6\cdot 15 \\ 1\cdot 63 & 3\cdot 79 \end{bmatrix}$$

and orthogonalization gives

$$U = \begin{bmatrix} 0\cdot 16 & 0\cdot 69 \\ 0\cdot 98 & -0\cdot 22 \\ 0\cdot 16 & 0\cdot 69 \end{bmatrix}.$$

The iteration proceeds as in the following table:

Iteration	U^T			B		μ_1, μ_2
1	1	0	0	4·0	1·0	10·16
	0	1	0	1·0	10·0	3·84
2	0·16	0·98	0·16	10·67	1·49	12·00
	0·69	−0·22	0·69	1·49	10·33	9·00
3	0·424	0·566	0·707	12·600	−0·431	12·657
	−0·251	0·824	−0·508	−0·431	9·391	9·334
4	0·421	0·454	0·785	12·677	−0·014	12·677
	−0·163	0·889	−0·427	−0·014	9·348	9·348
5	0·417	0·451	0·789	12·677	−0·0005	12·677
	−0·155	0·891	−0·427	−0·0005	9·348	9·348

We now have the required eigenvalues correct to three decimal places, although another two iterations are required to obtain the eigenvectors to the same accuracy, when we have

$$\mathbf{u}_1^T = [\ 0\cdot416, \quad 0\cdot450, \quad 0\cdot790]$$

and

$$\mathbf{u}_2^T = [-0\cdot153, \quad 0\cdot891, \quad -0\cdot427].$$

Exercises

5. Perform simultaneous iteration for λ_1 and λ_2 on the matrix

$$A = \begin{bmatrix} -12 & 11 & 1 \\ 11 & 8 & -2 \\ 1 & -2 & 12 \end{bmatrix},$$

taking

$$U = \begin{bmatrix} 1 & 0 \\ 0 & 1 \\ 0 & 0 \end{bmatrix}$$

initially.

6. Perform simultaneous iteration for λ_1 and λ_2 on the matrix

$$A = \begin{bmatrix} 1\cdot5 & 1\cdot0 & 0\cdot5 \\ 1\cdot0 & 1\cdot5 & 0\cdot25 \\ 0\cdot5 & 0\cdot25 & 2\cdot5 \end{bmatrix}.$$

6

Inverse iteration

6.1 Introduction

The algorithms for determining a few of the dominant/subdominant eigenvalues, which were outlined in the previous two chapters have, in the main, slow convergence rates. In this chapter, we turn to one of the most powerful methods available in connection with solving matrix eigenproblems. The technique, known as *inverse iteration*, is not only of general use in that it may be applied to the computation of an eigenvalue and/or an eigenvector, but also it possesses a fast rate of convergence. It is significant that we discuss this technique towards the middle of this book since it provides a strong connecting link between, on the one hand, the procedures for determining only a few eigenvalues/eigenvectors and, on the other hand, those methods which determine the complete set of eigenvalues/eigenvectors, or a significant subset of them. Frequently in the following chapters we will have occasion to refer to the technique of inverse iteration in connection with the determination of the eigenvectors of symmetric tridiagonal, or complex upper Hessenberg matrices. However we begin with a consideration of its application to determining eigenvalues.

6.2 Inverse iteration for an eigenvalue

Direct iteration of the form

$$\begin{aligned} \mathbf{y}^{(p+1)} &= B\mathbf{z}^{(p)} \\ \mathbf{z}^{(p+1)} &= \mathbf{y}^{(p+1)}/\|\mathbf{y}^{(p+1)}\|_\infty \end{aligned} \qquad (1)$$

gives under suitable conditions a convergent sequence of values approximating the dominant eigenvalue of B, and its associated eigenvector.

The process defined by

$$\begin{aligned} A\mathbf{y}^{(p+1)} &= \mathbf{z}^{(p)} \\ \mathbf{z}^{(p+1)} &= \mathbf{y}^{(p+1)}/\|\mathbf{y}^{(p+1)}\|_\infty \end{aligned} \qquad (2)$$

is equivalent to (1) with the matrix $B = A^{-1}$. Thus the sequence (2) will converge to the eigenvector corresponding to the eigenvalue of A of *smallest* modulus. To show this assume

$$\mathbf{z}^{(0)} = \sum_{i=1}^{n} \hat{\alpha}_i \mathbf{x}_i$$

where $A\mathbf{x}_i = \lambda_i \mathbf{x}_i$. Then (2) gives

$$\mathbf{z}^{(p)} = K_p \sum_{i=1}^{n} \hat{\alpha}_i \lambda_i^{-p} \mathbf{x}_i$$

where K_p is a scaling factor introduced by the rescaling part of (2). The vector $\mathbf{z}^{(p)}$ is 'richest' in the vector $\mathbf{x}_n$ corresponding to the smallest eigenvalue λ_n. Therefore the sequence $\mathbf{z}^{(p)}$ will tend to a multiple of $\mathbf{x}_n$ as $p \to \infty$, and also, for each j in general, $z_j^{(p+1)}/z_j^{(p)} \to 1/\lambda_n$ as $p \to \infty$. The process (2) is inverse iteration in its simplest form. Before we consider the computational details of an algorithm of the form (2) let us introduce a much more general, and useful version of it.

Consider a direct method using the matrix $(A-qI)^{-1}$ instead of A^{-1}. Corresponding to (2) we have (with renormalization)

$$\mathbf{z}^{(p)} = \sum_{i=1}^{n} \alpha_i (\lambda_i - q)^{-p} \mathbf{x}_i, \qquad \alpha_i = \hat{\alpha}_i K_p. \tag{3}$$

This vector will be dominated by the eigenvector $\mathbf{x}_d$ whose related eigenvalue is closest to the point q in the complex plane. (Notice the contrast with the result of shifting in direct iteration where only λ_1 or λ_n could be computed.)

This strategy of iterating directly (but in an efficient manner) with a matrix of the form $(A-qI)^{-1}$ constitutes the basis of inverse iteration. It is easily seen that given a good (or even a not so good) approximation to an eigenvalue the use of inverse iteration can provide rapid convergence to the eigenvalue and its associated eigenvector.

It might be thought that difficulty could be experienced if the initial choice of vector $\mathbf{z}^{(0)}$ was severely deficient in $\mathbf{x}_d$, but this is not the case. For suppose that

$$\lambda_d - q = \epsilon, \ |\lambda_i - q| > \theta \quad i \neq d,$$

then

$$\mathbf{z}^{(p)} = \alpha_d (\lambda_d - q)^{-p} \mathbf{x}_d + \sum_{\substack{i=1 \\ i \neq d}}^{n} \alpha_i (\lambda_i - q)^{-p} \mathbf{x}_i = \frac{\alpha_d}{\epsilon^p} \mathbf{x}_d + \sum_{\substack{i=1 \\ i \neq d}}^{n} \alpha_i (\lambda_i - q)^{-p} \mathbf{x}_i \tag{4}$$

from which it follows that

$$\frac{\epsilon^p}{\alpha_d} \mathbf{z}^{(p)} - \mathbf{x}_d = \frac{\epsilon^p}{\alpha_d} \sum_{\substack{i=1 \\ i \neq d}}^{n} \alpha_i (\lambda_i - q)^{-p} \mathbf{x}_i$$

or that

$$\left\| \frac{\epsilon^p}{\alpha_d} \mathbf{z}^{(p)} - \mathbf{x}_d \right\| \leqslant \frac{\epsilon^p}{|\alpha_d|} \sum_{\substack{i=1 \\ i \neq d}}^{n} |\alpha_i| \, \theta^{-p} = \left(\frac{\epsilon}{\theta}\right)^p \frac{1}{|\alpha_d|} \sum_{\substack{i=1 \\ i \neq d}}^{n} |\alpha_i|. \qquad (5)$$

Since we are seeking convergence to λ_d we must have chosen q so that λ_d is the eigenvalue closest to q. Thus $\epsilon < \theta$ and unless there is a cluster of eigenvalues close to q, $\epsilon \ll \theta$. Thus, irrespective of α_d the convergence rate is proportional to ϵ/θ. The possible difficulties with close or coincident eigenvalues will be discussed later.

We have so far considered only the theoretical aspects of inverse iteration. To take advantage of its power, it must be implemented efficiently. This we discuss in the next section.

6.3 Computational procedure for inverse iteration

The (sequence of) vectors generated by inverse iteration are all solutions of the set of linear equations

$$(A - qI)\,\mathbf{y}^{(p+1)} = \mathbf{z}^{(p)} \qquad (6)$$

for some choice of q. Notice that in the computation of the eigenvalue closest to q the matrix $(A - qI)$ is the same throughout the iteration, and so the process amounts to the successive solution of the same linear system with different right-hand sides. Thus we see that considerable economy can be achieved by constructing the LU decomposition of $(A - qI)$ only once. (It will of course be necessary to introduce interchanges as discussed in Chapter 3; we omit them for simplicity of presentation.) Further it will frequently be the case that A is either a tridiagonal or an upper Hessenberg matrix, and in this situation even greater economy of storage and computation can be achieved. Thus, ignoring interchanges, we will be using the process

$$\begin{aligned} L\mathbf{z} &= \mathbf{z}^{(p)} \\ U\mathbf{y}^{(p+1)} &= \mathbf{z}, \end{aligned} \qquad (7)$$

where

$$LU = A - qI.$$

We have noticed that the choice of initial vector is not particularly important. In fact, the standard procedure is to omit the first step in (7) for $p = 0$ and to obtain our first iterate by the relation

$$U\mathbf{y}^{(1)} = \mathbf{e} \qquad (8)$$

where $\mathbf{e}^T = (1, 1, \ldots, 1)$. This amounts to choosing

$$\mathbf{y}^{(0)} = L\mathbf{e}.$$

The above strategy is the recommended starting procedure and it has been found to work extremely well in practice. Moreover in the case when we are using inverse iteration to determine the eigenvector associated with an approximation q to an eigenvalue, the convergence is extremely rapid, occurring in one or two iterations.

The fast convergence may be impaired if there is more than one eigenvalue close to q. From equation (4) we can see that the successive iterates, whether normalized or not, will contain in addition to $\mathbf{x}_d$, a substantial multiple of the eigenvector associated with the close eigenvalue. A simple example demonstrating this is given by the diagonal matrix

$$(A-qI) = \begin{bmatrix} \epsilon & 0 & 0 \\ 0 & \epsilon+\delta & 0 \\ 0 & 0 & 1 \end{bmatrix}$$

where ϵ is a small number. The explicit expression for $(A-qI)^{-1}$ is given by

$$\begin{bmatrix} \frac{1}{\epsilon} & 0 & 0 \\ 0 & \frac{1}{\epsilon+\delta} & 0 \\ 0 & 0 & 1 \end{bmatrix}$$

and it follows that the unnormalized sequence ($\mathbf{z}^{(p)} = \mathbf{y}^{(p)}$) produced by inverse iteration gives

$$\mathbf{y}^{(p)} = \begin{bmatrix} \left(\frac{1}{\epsilon}\right)^p & 0 & 0 \\ 0 & \left(\frac{1}{\epsilon+\delta}\right)^p & 0 \\ 0 & 0 & 1 \end{bmatrix} \mathbf{y}^{(0)}.$$

Since ϵ is small and δ, the separation of the two eigenvalues, may be small or large, we see that the $\mathbf{y}^{(p)}$ may be tending to one of two limits:

(*a*) $\epsilon \ll \delta$, $\quad \mathbf{y}^{(\infty)} \propto (1, 0, 0)^T$

(*b*) $\epsilon = 0(\delta)$, $\quad \mathbf{y}^{(\infty)} \propto (1, \beta, 0)^T$,

where $\beta = 0(1)$. In case (*b*), we have two close eigenvalues and the resulting vector is rich in both eigenvectors. Of course, a similar situation would result if we used $(A-(q-\delta)\,I)^{-1}$. The two vectors thus produced would both be rich in the eigenvectors associated with the two close eigenvalues. More effort would be required to obtain the respective eigenvalue/vector pair.

The above type of procedure may be used to advantage when the matrix has two coincident eigenvalues. It is then obvious that inverse iteration as defined above can only produce one eigenvector. (Of course if the matrix has non-linear elementary divisors this may be a true representation of the situation.) If there is more than one eigenvector associated with a multiple eigenvalue (for example if the matrix is symmetric) we can obtain information on the subspace by perturbing the eigenvalue q slightly. To illustrate the technique, consider the diagonal matrix

$$\begin{bmatrix} 1 & 0 & 0 \\ 0 & 1 & 0 \\ 0 & 0 & 2 \end{bmatrix}$$

and suppose an approximate double eigenvalue $q = 1-\epsilon$ is known. Then taking $q = 1-\epsilon$,

$$(A-qI) = \begin{bmatrix} \epsilon & 0 & 0 \\ 0 & \epsilon & 0 \\ 0 & 0 & 1+\epsilon \end{bmatrix}$$

and

$$(A-qI)^{-1} = \begin{bmatrix} \frac{1}{\epsilon} & 0 & 0 \\ 0 & \frac{1}{\epsilon} & 0 \\ 0 & 0 & \frac{1}{1+\epsilon} \end{bmatrix}$$

and a sequence of inverse iterates would produce a vector proportional to $(1, 1, 0)^T$. Obviously, a simple perturbation of the approximate eigenvalue $(1-\epsilon)$ will provide the same eigenvector. Therefore, it may be necessary to perturb the matrix slightly to obtain a second different eigenvector. If we introduce an element δ in the (2, 3) and (3, 2) positions we have that

$$(A-qI) = \begin{bmatrix} \epsilon & 0 & 0 \\ 0 & \epsilon & \delta \\ 0 & \delta & 1+\epsilon \end{bmatrix}$$

and

$$(A-qI)^{-1} = \begin{bmatrix} \frac{1}{\epsilon} & 0 & 0 \\ 0 & \frac{1+\epsilon}{\epsilon^2+\epsilon-\delta^2} & \frac{-\delta}{\epsilon^2+\epsilon-\delta^2} \\ 0 & \frac{-\delta}{\epsilon^2+\epsilon-\delta^2} & \frac{\epsilon}{\epsilon^2+\epsilon-\delta^2} \end{bmatrix}$$

and inverse iteration now produces a vector of the form $(1, \beta, 0)^T$ where $\beta \neq 1$. Thus we have full information on the subspace of eigenvectors associated with the double eigenvalue 1, from the two vectors $(1, 1, 0)^T$, $(1, \beta, 0)^T$ where $\beta \neq 1$.

Notice that the above difficulty should not occur when dealing with symmetric tridiagonal matrices. If a symmetric tridiagonal matrix T has a multiple eigenvalue then one or more of the off-diagonal elements must be zero (see Section 9.2), and we would work with matrices of lower order. In practice, a reduction by the Givens or Householder processes (see Chapter 8) may produce a tridiagonal matrix with zero or very small off-diagonal quantities, but this need not imply that the matrix has multiple or even close eigenvalues. In general, it will be sufficient simply to perturb the value of the eigenvalue in order to obtain more information on the eigenvector subspace.

Finally let us consider what happens when we have a multiple eigenvalue with an associated non-linear elementary divisor. As an example consider the simple 2×2 Jordan submatrix

$$A = \begin{bmatrix} 2 & 1 \\ 0 & 2 \end{bmatrix}.$$

Suppose $(A-qI)$, where $q = 2-\epsilon$, is used for inverse iteration. Then

$$(A-qI)^{-1} = \begin{bmatrix} \frac{1}{\epsilon} & -\frac{1}{\epsilon^2} \\ 0 & \frac{1}{\epsilon} \end{bmatrix}$$

and it follows that $\mathbf{y}^{(\infty)} \propto (1, 0)^T$, the required eigenvector. Notice that the presence of a non-linear divisor has not impeded the inverse iteration. Note further that perturbation of the matrix in the manner outlined above would not lead to a second eigenvector.

There is little doubt that the technique of inverse iteration is one of the most powerful for finding an eigenvector associated with an eigenvalue which is known only approximately. It is also efficient in finding a particular eigenvalue which satisfies some criterion, for example being nearest to a point in the complex plane.

Exercises

1. Use inverse iteration to find the eigenvalue nearest to six and corresponding eigenvector of the matrix

$$\begin{bmatrix} 6 & 2 & 1 \\ 2 & 3 & 1 \\ 1 & 1 & 1 \end{bmatrix}.$$

2. What happens to the inverse iteration process if the matrix has a complex conjugate pair of eigenvalues $a \pm ib$ and we use a real $q = a$? (Assume all other eigenvalues lie well outside the disk $(x-a)^2 + y^2 \leqslant b^2$.)

3. Find the eigenvectors associated with the eigenvalue four of the matrix

$$\begin{bmatrix} 4 & 0 & 0 \\ 0 & 3 & 1 \\ 0 & 1 & 3 \end{bmatrix}.$$

4. If we start the inverse iteration process with a very good approximation to the eigenvalue, then the matrix $(A - qI)$ is nearly singular. This would appear to cause trouble in implementing the algorithm, but in practice no difficulty arises. Why?

7

Jacobi's method

7.1 Introduction

In the preceding three chapters, several algorithms were introduced which computed a single eigenvalue and the corresponding eigenvector at any stage. It was further shown how such a process could successively be used to compute the complete eigensystem of a matrix. In this and subsequent chapters, we turn to those algorithms which are designed to compute the complete eigensystem from the outset. The basis of these algorithms will be the elementary similarity transformations introduced in Chapter 3, and our aim will be to choose these transformations in such a way that we reduce the original matrix A to one whose eigensystem is more easily obtained. In particular, for Hermitian matrices A, we know that there exists a unitary matrix H such that,

$$HAH^* = D,$$

where D is a real diagonal matrix, and in this chapter, we consider how a Hermitian matrix may be reduced to diagonal form.

It seems natural to attempt to construct the matrix H by using successive elementary unitary transformations. In fact, we construct an approximation to H which is such that HAH^* has off-diagonal elements which are negligible in comparison to the diagonal elements. Thus our algorithm must effect a reduction in the off-diagonal elements of the matrix A.

Since A is Hermitian it follows that the matrix $B = HAH^*$ has the same Euclidean norm as A. In other words

$$\|B\|_E^2 = \|A\|_E^2$$

where

$$\|A\|_E^2 = \sum_{i=1}^{n} \sum_{j=1}^{n} |a_{ij}|^2.$$

This may easily be proved by noticing that

$$\begin{aligned}\|A\|_E^2 &= \text{trace } (A^*A)\\ &= \text{trace } (A^2)\\ &= \sum_{i=1}^{n} \lambda_i^2\end{aligned}$$

where λ_i is an eigenvalue of A. Since a similarity transformation leaves the eigenvalues unaltered it follows that

$$\|B\|_E^2 = \|A\|_E^2.$$

Therefore if we construct a sequence of matrices A_s where A_{s+1} is obtained from A_s by an elementary unitary similarity transformation, we have the result that the Euclidean norm of each matrix A_s is equal to the Euclidean norm of $A_1 = A$.

But our aim is to reduce the size of the off-diagonal elements of A by means of the transformations. Therefore if we define

$$E_{(s)} = \sum_{\substack{i,j=1\\ i\neq j}}^{n} |a_{ij}^{(s)}|^2 \tag{1}$$

$$D_{(s)} = \sum_{i=1}^{n} |a_{ii}^{(s)}|^2 \tag{2}$$

and if we choose our parameters so that

$$E_{(s+1)} \leqslant E_{(s)}$$

we have that

$$D_{(s+1)} \geqslant D_{(s)}.$$

The most rapid convergence of the above algorithm to diagonal form would obviously be obtained if we maximized the quantity $D_{(s+1)} - D_{(s)}$ at each stage. In a later section, we will in fact see that this is achieved by choosing our parameters according to a simple rule. This rule, and hence the algorithm which bears his name, were first given by C. G. J. Jacobi in 1846. For ease of presentation, we will carry out the analysis for a real symmetric matrix A. The matrix H is then orthogonal. The extension to the more general case when A is Hermitian is straightforward.

7.2 Jacobi's algorithm

This classical algorithm aims to reduce the matrix A_1 to diagonal form by carrying out a sequence of plane rotations. Thus we construct successively the sequence of matrices A_s such that

$$A_{s+1} = R(p, q)\, A_s R^T(p, q)$$

where $R(p, q)$ is a rotation through an angle θ in the (p, q) plane. The strategy used in choosing the plane and in determining the rotation angle θ is straightforward. We search through the elements of A_s lying above the main diagonal to determine the entry $a_{pq}^{(s)}$ of maximum modulus. (We need only carry through the search in the upper part of A by symmetry considerations.) Our rotation angle θ is then chosen to ensure that $a_{pq}^{(s+1)}$ is identically zero. In Chapter 3, it was shown that a similarity transformation using the plane rotation $R(p, q)$ affected the entries in rows and columns p and q only. The modified elements are given by

$$
\begin{aligned}
a_{ip}^{(s+1)} &= a_{ip}^{(s)} \cos \theta + a_{iq}^{(s)} \sin \theta = a_{pi}^{(s+1)} \\
a_{iq}^{(s+1)} &= -a_{ip}^{(s)} \sin \theta + a_{iq}^{(s)} \cos \theta = a_{qi}^{(s+1)} \qquad i \neq p, q \\
a_{pp}^{(s+1)} &= a_{pp}^{(s)} \cos^2 \theta + 2a_{pq}^{(s)} \cos \theta \sin \theta + a_{qq}^{(s)} \sin^2 \theta \qquad (3) \\
a_{qq}^{(s+1)} &= a_{pp}^{(s)} \sin^2 \theta - 2a_{pq}^{(s)} \cos \theta \sin \theta + a_{qq}^{(s)} \cos^2 \theta \\
a_{pq}^{(s+1)} &= (a_{qq}^{(s)} - a_{pp}^{(s)}) \cos \theta \sin \theta + a_{pq}^{(s)} (\cos^2 \theta - \sin^2 \theta) \\
&= a_{qp}^{(s+1)}.
\end{aligned}
$$

If the rotation angle θ is chosen to annihilate $a_{pq}^{(s+1)}$ we require

$$-(a_{pp}^{(s)} - a_{qq}^{(s)}) \sin 2\theta + 2a_{pq}^{(s)} \cos 2\theta = 0,$$

that is,

$$\tan 2\theta = \frac{2a_{pq}^{(s)}}{(a_{pp}^{(s)} - a_{qq}^{(s)})}. \qquad (4)$$

The angle θ is chosen so that

$$-\frac{\pi}{4} \leqslant \theta \leqslant \frac{\pi}{4}.$$

Moreover if $a_{pp}^{(s)} - a_{qq}^{(s)} = 0$ then θ is chosen to be $(a_{pq}^{(s)}/|a_{pq}^{(s)}| . \pi/4)$. Notice that if $a_{pq}^{(s)} = 0$ then no rotation would be required. (Of course under our present strategy, if $a_{pq}^{(s)} = 0$ then A_s would already be diagonal.)

From equations (3) it is easily proved that

$$E_{(s+1)} = E_{(s)} - 2(a_{pq}^{(s)})^2, \qquad (5)$$

giving

$$D_{(s+1)} = D_{(s)} + 2(a_{pq}^{(s)})^2,$$

verifying that the sequence of matrices A_s is indeed tending to diagonal form. Thus ultimately the matrix A_s will effectively be a diagonal matrix. However we must also show that the sequence tends to a *fixed* diagonal matrix. In other words, we must show that

$$|a_{pp}^{(s+1)} - a_{pp}^{(s)}| \to 0$$

as $s \to \infty$. Now from (3) we have that

$$
\begin{aligned}
a_{pp}^{(s+1)} - a_{pp}^{(s)} &= -a_{pp}^{(s)}(1 - \cos^2 \theta) + 2a_{pq}^{(s)} \cos \theta \sin \theta + a_{qq}^{(s)} \sin^2 \theta \\
&= (a_{qq}^{(s)} - a_{pp}^{(s)}) \sin^2 \theta + 2a_{pq}^{(s)} \cos \theta \sin \theta.
\end{aligned}
$$

If $a_{qq}^{(s)}-a_{pp}^{(s)} = 0$ we have immediately that

$$|a_{pp}^{(s+1)}-a_{pp}^{(s)}| = |a_{pq}^{(s)}|$$

since $|\theta| = \pi/4$. If $a_{qq}^{(s)}-a_{pp}^{(s)} \neq 0$ then using equation (4) and straightforward trigonometric manipulation we find

$$|a_{pp}^{(s+1)}-a_{pp}^{(s)}| = |a_{pq}^{(s)}|\,|\tan\theta|.$$

However $|\theta| \leqslant \pi/4$ and thus in general

$$|a_{pp}^{(s+1)}-a_{pp}^{(s)}| \leqslant |a_{pq}^{(s)}|.$$

If we have reached the stage when $|a_{pq}^{(s)}| \leqslant \epsilon$ it follows that

$$|a_{pp}^{(s+1)}-a_{pp}^{(s)}| \leqslant \epsilon,$$

and likewise

$$|a_{qq}^{(s+1)}-a_{qq}^{(s)}| \leqslant \epsilon.$$

Therefore the Jacobi algorithm does generate a sequence of matrices which tend to a *fixed* diagonal matrix, which is similar to our initial matrix. Reference to Section 2.9 shows that the process is a stable one.

Example *Apply the Jacobi method to the matrix*

$$A = \begin{bmatrix} 1 & 0 & 2 \\ 0 & 2 & 1 \\ 2 & 1 & 1 \end{bmatrix}.$$

Step 1. Let $A_1 = A$. Then $p=1$, $q=3$ and $\theta = \pi/4$. The plane rotation matrix is

$$R(1,3) = \begin{bmatrix} 0\cdot7071 & 0 & 0\cdot7071 \\ 0 & 1 & 0 \\ -0\cdot7071 & 0 & 0\cdot7071 \end{bmatrix},$$

and

$$A_2 = \begin{bmatrix} 3 & 0\cdot7071 & 0 \\ 0\cdot7071 & 2 & 0\cdot7071 \\ 0 & 0\cdot7071 & -1 \end{bmatrix}.$$

Step 2. $p = 1, q = 2$; $\tan 2\theta = 1\cdot4142$,

$$R(1,2) = \begin{bmatrix} 0\cdot8881 & 0\cdot4596 & 0 \\ -0\cdot4596 & 0\cdot8881 & 0 \\ 0 & 0 & 1 \end{bmatrix},$$

$$A_3 = \begin{bmatrix} 3\cdot3659 & 0 & 0\cdot3250 \\ 0 & 1\cdot6339 & 0\cdot6280 \\ 0\cdot3250 & 0\cdot6280 & -1 \end{bmatrix}.$$

Step 3. $p = 2, q = 3$; $\tan 2\theta = 0 \cdot 4768$,

$$R(2, 3) = \begin{bmatrix} 1 & 0 & 0 \\ 0 & 0\cdot9753 & 0\cdot2263 \\ 0 & -0\cdot2263 & 0\cdot9753 \end{bmatrix},$$

$$A_4 = \begin{bmatrix} 3\cdot3659 & 0\cdot0735 & 0\cdot3170 \\ 0\cdot0735 & 1\cdot7801 & 0 \\ 0\cdot3170 & 0 & -1\cdot1447 \end{bmatrix}.$$

Step 4. $p = 1, q = 3$; $\tan 2\theta = 0 \cdot 1405$

$$R(1, 3) = \begin{bmatrix} 0\cdot9976 & 0 & 0\cdot0698 \\ 0 & 1 & 0 \\ -0\cdot0698 & 0 & 0\cdot9976 \end{bmatrix},$$

$$A_5 = \begin{bmatrix} 3\cdot3883 & 0\cdot0733 & 0 \\ 0\cdot0733 & 1\cdot7801 & 0\cdot0051 \\ 0 & 0\cdot0051 & -1\cdot1670 \end{bmatrix}.$$

The completion of this example is left as an exercise.

Exercise

1. Prove that $E_{(s+1)} \leqslant (1-2/n(n-1))\, E_{(s)}$ for an $n \times n$ symmetric matrix A.

7.3 Variants of the Jacobi algorithm

In the form given above, the implementation of the Jacobi algorithm on a computer is somewhat time consuming. This arises primarily from the need to search at each stage for the element a_{pq} of maximum modulus. There are two alternative strategies which are more acceptable in practice.

The first is the *cyclic Jacobi method* in which no searches are carried out. The elements are annihilated according to a cyclic ordering of the elements of the matrix. The usual ordering is to choose the rotations to annihilate the elements in positions $(1, 2)$, $(1, 3), \ldots, (1, n)$, $(2, 3), \ldots, (n-1, n)$, that is rowwise. At each stage a check is introduced to test whether the element about to be reduced to zero is of significant size or not, in comparison to the sum of squares of the diagonal elements.

The second strategy is the *threshold Jacobi method* in which limited searching takes place. At each stage of the algorithm a threshold is set and a search is carried out to find an element which exceeds the threshold in modulus. This element is then annihilated and the process repeated. Since $E_{(s)}$, the sum of the squares of off-diagonal elements, is tending to zero, it follows that the level of the threshold will require to be reduced as the algorithm progresses.

Eventually a threshold will require to be set which is close to the smallest number representable on the computer being used and the algorithm may be assumed to have converged.

It may be shown that for the above three Jacobi strategies the convergence to diagonal form is ultimately quadratic in that

$$E_{(s+N)} \leqslant K\{E_{(s)}\}^2,$$

where K depends on the separation of the eigenvalues and on the order n of A and where $N = \frac{1}{2}n(n-1)$.

Exercise

2. Show that a complete sweep of the cyclic Jacobi method applied to a 3×3 matrix with off-diagonal elements $0(\epsilon)$ gives a matrix with off-diagonal elements $0(\epsilon^2)$.

7.4 The maximizing property of the classical Jacobi algorithm

The choice of rotation angle in each of the above Jacobi procedures is a natural one leading to an algorithm which does converge to a diagonal matrix. This choice ensured that

$$D_{(k+1)} - D_{(k)} \geqslant 0$$

where

$$D_{(k)} = \sum_{i=1}^{n} (a_{ii}^{(k)})^2.$$

The most rapid convergence would obviously be achieved if at each stage we chose the plane of rotation (p, q) and the angle θ in order to maximize

$$(D_{(k+1)} - D_{(k)}).$$

In this section, we will show that the classical Jacobi algorithm does, in fact, achieve this very goal. From equations (3) we may derive, after a little manipulation, the result

$$D_{(k+1)} - D_{(k)} = -2cs[c(a_{pp}^{(k)} - a_{qq}^{(k)}) + 2sa_{pq}^{(k)}][s(a_{pp}^{(k)} - a_{qq}^{(k)}) - 2ca_{pq}^{(k)}] \tag{6}$$

where $c \equiv \cos\theta$, $s \equiv \sin\theta$. Consider the maximization of $D_{(k+1)} - D_{(k)}$ with respect to the rotation angle θ. Elementary differentiation yields the result

$$\frac{\mathrm{d}}{\mathrm{d}\theta}(D_{(k+1)} - D_{(k)}) = -4(a_{pp}^{(k)} - a_{qq}^{(k)})^2[cs - X(c^2 - s^2)][4csX + (c^2 - s^2)]$$

where $X = a_{pq}^{(k)}/(a_{pp}^{(k)} - a_{qq}^{(k)})$. Assuming $a_{pp}^{(k)} - a_{qq}^{(k)} \neq 0$ we note that $D_{(k+1)} - D_{(k)}$

has two turning values. Computation of $d^2/d\theta^2\,(D_{(k+1)}-D_{(k)})$ verifies that $D_{(k+1)}-D_{(k)}$ has a maximum when

$$cs = X(c^2-s^2)$$

that is when

$$\tan 2\theta = \frac{2a_{pq}^{(k)}}{(a_{pp}^{(k)}-a_{qq}^{(k)})},$$

which is precisely the Jacobi strategy. The case $a_{pp}^{(k)}-a_{qq}^{(k)} = 0$ may be readily dealt with by considering equation (6) which then becomes

$$D_{(k+1)}-D_{(k)} = 2(a_{pq}^{(k)})^2 \sin^2 2\theta$$

giving a maximum at $|\theta| = \pi/4$.

It follows that of all the diagonalizing Jacobi algorithms which have $D_{(k+1)} \geqslant D_{(k)}$, the one which maximizes the left-hand side of (6) is the classical Jacobi algorithm. Notice that the various operating strategies, classical, cyclic or threshold serve to optimize or nearly optimize $D_{(k+1)}-D_{(k)}$ with respect to the plane of rotation. This result is immediate, for if we optimize with respect to θ for fixed (p, q) we obtain

$$D_{(k+1)}-D_{(k)} = 2(a_{pq}^{(k)})^2.$$

7.5 Calculation of the eigenvectors

Using any of the above Jacobi procedures we eventually obtain a matrix which we take to be diagonal to a given precision. It follows that we have

$$RAR^T = D$$

where R is a product of plane rotation matrices.
Since

$$AR^T = R^T D$$

it follows that the columns of R^T give the eigenvectors of the matrix A.

There are two distinct processes used depending on whether all or only some of the eigenvectors are to be obtained. We assume that the algorithm is being implemented on a computer. Then if only the eigenvector related to the jth eigenvalue is required then we need only store the plane (p, q) of each rotation and $\cos\theta$ and $\sin\theta$ for this rotation. [Note however that as several sweeps, say s, of the upper triangle of A will in general be required then a total of $2sn(n-1)$ storage locations will be used.] Then the jth eigenvector may be obtained by multiplying $\mathbf{e}_j$ by the rotations in the correct order. However if many eigenvectors are required or if storage is at a premium then it is more economical to compute R^T as the rotations are determined. This may be more expensive on computing time but requires only n^2 storage locations.

It should be observed that the eigenvectors are determined simultaneously with the eigenvalues in the Jacobi process. Moreover the eigenvectors given by R^T do form an orthogonal set. However if A has eigenvalues close to one another the eigenvectors may not be obtained very accurately although they will still form an orthogonal set. The fact that we always obtain a full set of eigenvectors by Jacobi's method gives considerable weight to its use. However as we shall see there are several faster techniques, especially when the eigenvalues are distinct.

8

Givens' and Householder's methods

8.1 Introduction

The algorithm of Jacobi, discussed in the preceding chapter, is a very reliable process for the determination of the complete eigensystem of a Hermitian matrix. However it should be borne in mind that at each stage in the reduction to diagonal form one must work with the complete matrix (though advantage can be taken of symmetry). Elements which are reduced to zero in one step may become non-zero, and significant, in later steps, and thus the algorithm is an infinite process which can be slow to converge, particularly for larger matrices. In addition, we frequently do not want to calculate the eigenvectors of the matrix, but only the eigenvalues.

We now study algorithms which serve only to reduce the Hermitian matrix A to a form whose eigenproblem is more simply tackled. The particular form to which reduction is made is that of a real symmetric tridiagonal matrix, and an important feature of this type of algorithm is that the reduction may be accomplished in a finite number of steps.

The techniques for finding the eigensystem of a symmetric tridiagonal matrix will be studied in Chapters 9 and 10.

8.2 Givens' method

In Section 3.7, the use of elementary unitary matrices was discussed in connection with the elementary similarity (unitary) transformation of a matrix. The use of plane rotations in this context to reduce a Hermitian matrix to tridiagonal form constitutes a method developed by W. J. Givens in 1954. We will carry out the analysis for a real, symmetric matrix.

The angle of rotation is chosen to annihilate an element which has been modified only by premultiplication by $R(j, k)$ or (by symmetry) only by post-multiplication by $R^T(j, k)$. In terms of the display in Section 3.7 for the case $j = 2, k = 4$ (shown again below for convenience), we annihilate an element marked 'a' or 'b' but *not* 'c'. (The Jacobi strategy corresponded to the annihilation of an off-diagonal element 'c'.) The planes of rotation are also chosen in a particular order. This sequence is such that an element reduced to zero by a plane rotation never becomes non-zero in a later transformation. In this sense, the algorithm has connections with the decomposition outlined in Section 3.6.

$$\begin{bmatrix} x & b & x & b & x \\ a & c & a & c & a \\ x & b & x & b & x \\ a & c & a & c & a \\ x & b & x & b & x \end{bmatrix}$$

The precise ordering of rotations is as follows. Starting with the first column we annihilate elements in positions (3, 1) to $(n, 1)$ by rotations in planes (2, 3), . . . , $(2, n)$. A similar strategy is applied columnwise across the matrix. Thus, in general, the transformation on the jth column reduces the elements $(j+2, j)$ to (n, j) to zero by transformations in planes $(j+1, j+2)$ to $(j+1, n)$. By symmetry an element in position (k, j) is annihilated when an element in position (j, k) is reduced to zero. It may be readily verified that the above ordering of transformations is such that a zero introduced by one transformation is not affected by subsequent transformations. As an example consider the following situation:

$$\begin{bmatrix} x & x & 0 & 0 & 0 \\ x & x & x & 0 & x \\ 0 & x & x & x & x \\ 0 & 0 & x & x & x \\ 0 & x & x & x & x \end{bmatrix}.$$

We employ a rotation in the plane (3, 5) to reduce the element (5, 2) to zero. The transformation is completed by postmultiplication by $R^T(3, 5)$. Only elements in rows and columns 3 and 5 are affected, and in the pre-(post-) multiplication rows (columns) are treated independently. It follows that we may choose our angle of rotation to annihilate element (5, 2) without modifying elements (3, 1) and (5, 1). In general, the element $a_{jk}^{(s+1)}$ is given by

$$a_{jk}^{(s+1)} = -a_{j,j+1}^{(s)} \sin \theta + a_{jk}^{(s)} \cos \theta \tag{1}$$

(see equations 3, Chapter 7). From this equation we see that $a_{jk}^{(s+1)}$ will be zero if

$$\tan \theta = \frac{a_{jk}^{(s)}}{a_{j,j+1}^{(s)}}.$$

It follows that we may transform the symmetric matrix A_1 through a sequence of matrices A_s so that A_M is symmetric tridiagonal, where $M \leqslant \frac{1}{2}(n-1)(n-2)$. Moreover it may be calculated that approximately $\frac{4}{3}n^3$ multiplications are required to achieve the reduction to tridiagonal form. (In comparison, a single sweep of the Jacobi process requires $2n^3$ operations.) Also it must be noted that each rotation requires the computation of a square root, and thus the transformation requires in addition the calculation of $\frac{1}{2}(n-1)(n-2)$ square roots.

Example *Apply Givens' method to tridiagonalize the matrix*

$$A = \begin{bmatrix} 1 & 2 & 1 & 2 \\ 2 & 2 & -1 & 1 \\ 1 & -1 & 1 & 1 \\ 2 & 1 & 1 & 1 \end{bmatrix}.$$

Step 1. Let $A_1 = A$.

$$p = 2, q = 3; \tan\theta = a_{13}/a_{12} = \tfrac{1}{2}$$

$$R(2, 3) = \begin{bmatrix} 1 & 0 & 0 & 0 \\ 0 & 0{\cdot}8944 & 0{\cdot}4472 & 0 \\ 0 & -0{\cdot}4472 & 0{\cdot}8944 & 0 \\ 0 & 0 & 0 & 1 \end{bmatrix},$$

$$A_2 = \begin{bmatrix} 1 & 2{\cdot}2361 & 0 & 2 \\ 2{\cdot}2361 & 1 & -1 & 1{\cdot}3416 \\ 0 & -1 & 2 & 0{\cdot}4472 \\ 2 & 1{\cdot}3416 & 0{\cdot}4472 & 1 \end{bmatrix}.$$

Step 2. $p = 2, q = 4; \tan\theta = a_{14}/a_{12} = 0{\cdot}8944$

$$A_3 = \begin{bmatrix} 1 & 3 & 0 & 0 \\ 3 & 2{\cdot}3333 & -0{\cdot}4472 & 0{\cdot}1491 \\ 0 & -0{\cdot}4472 & 2 & 1 \\ 0 & 0{\cdot}1491 & 1 & -0{\cdot}3333 \end{bmatrix}.$$

Step 3. $p = 3, q = 4; \tan\theta = a_{24}/a_{23} = -0{\cdot}3333$

$$A_4 = \begin{bmatrix} 1 & 3 & 0 & 0 \\ 3 & 2{\cdot}3333 & -0{\cdot}4714 & 0 \\ 0 & -0{\cdot}4714 & 1{\cdot}1667 & 1{\cdot}5000 \\ 0 & 0 & 1{\cdot}5000 & 0{\cdot}5000 \end{bmatrix}.$$

Exercises

1. Verify that approximately $\frac{4}{3}n^3$ multiplications are required in the Givens reduction.

2. Use Givens' process to reduce to tridiagonal form the matrix

$$\begin{bmatrix} 0 & 12 & 16 & -15 \\ 12 & 288 & 309 & 185 \\ 16 & 309 & 312 & 80 \\ -15 & 185 & 80 & -600 \end{bmatrix}.$$

(All arithmetic involved may be carried through easily. The resulting matrix has zeros on the diagonal and the elements 25, 625, 125 respectively down the sub-diagonal.)

3. Show that Givens' process applied to a Hermitian matrix will produce a *real symmetric* tridiagonal matrix.

8.3 Householder's method

Although the Givens procedure gives a satisfactory reduction to tridiagonal form for a symmetric (or, more generally, Hermitian) matrix, a more efficient process based on the use of elementary Hermitian matrices (see Sections 3.4 and 3.7) in place of the plane rotations was suggested by A. S. Householder in 1958. We achieve the reduction to simplified form by means of $(n-2)$ elementary Hermitian transformations requiring approximately half the number of multiplications required by Givens' method and requiring only $(n-2)$ square roots to be calculated. Again, we will carry out the analysis for a real, symmetric matrix. However, it is convenient to retain the general unitary property of the transforming matrices.

A unitary transformation of the matrix A by the elementary Hermitian matrix

$$P^{(r)} = I-2\mathbf{w}^{(r)}\{\mathbf{w}^{(r)}\}^* \tag{2}$$

where

$$\{\mathbf{w}^{(r)}\}^T = (0, \ldots, 0, w_{r+1}^{(r)}, \ldots, w_n^{(r)}) \tag{3}$$

and

$$\{\mathbf{w}^{(r)}\}^* \mathbf{w}^{(r)} = 1 \tag{4}$$

modifies the elements in rows $r+1$ to n and columns $r+1$ to n of the matrix A. Consider a 5×5 matrix A transformed by $P^{(1)}$. The configuration is given by

$$\begin{bmatrix} x & b & b & b & b \\ a & c & c & c & c \\ a & c & c & c & c \\ a & c & c & c & c \\ a & c & c & c & c \end{bmatrix}$$

where elements marked

- a have been modified only in the premultiplication of A by $P^{(1)}$
- b have been modified only in the postmultiplication of A by $P^{(1)}(=P^{(1)*})$
- c have been modified by both the pre- and postmultiplications
- x are unchanged.

By the symmetry of A the matrix $P^{(1)}AP^{(1)}$ is also symmetric. We may choose $P^{(1)}$ so that the elements (3, 1) to (5, 1) of $P^{(1)}AP^{(1)}$ are zero (see Section 3.4). Let $A_2 = P^{(1)}AP^{(1)}$, and consider now the matrix $P^{(2)}A_2P^{(2)}$. This has the form

$$A_3 = \begin{bmatrix} x & x & b & b & b \\ x & x & b & b & b \\ a & a & c & c & c \\ a & a & c & c & c \\ a & a & c & c & c \end{bmatrix}.$$

Since the premultiplication of A_2 by $P^{(2)}$ treats columns independently it follows that the elements marked 'a' in the first column of A_3 are still zero. Therefore we may choose $P^{(2)}$ in order to annihilate elements (4, 2), (5, 2) without modifying zeros in the first column (and row). The reduction is completed in one more step by the annihilation of the (5, 3) element by the elementary Hermitian matrix $P^{(3)}$.

The determination of the vector $\mathbf{w}^{(r)}$ is given by equations (2) and (3), Chapter 3. For completeness we give an equivalent formulation here. Let $\mathbf{x}$ be the rth column of the matrix A_r. We require to choose $\mathbf{w}^{(r)}$ so that the $r+2$ to n elements of the rth column of A_{r+1}, where $A_{r+1} = P^{(r)}A_rP^{(r)}$, are zero. We partition $\mathbf{x}$ in the following way

$$\mathbf{x}^T = (\hat{\mathbf{x}}^T, x_{r+1}, \mathbf{y}^T)$$

where $\hat{\mathbf{x}}$ is $r \times 1$, $\mathbf{y}$ is $\{n-(r+1)\} \times 1$. Corresponding to this, we let

$$\mathbf{w}^{(r)T} = (\mathbf{0}, w_{r+1}^{(r)}, \mathbf{u}^T).$$

To determine $\mathbf{w}^{(r)}$ we only need to consider the vector $P^{(r)}\mathbf{x}$. This is given by

$$\begin{bmatrix} \hat{\mathbf{x}} \\ x_{r+1} \\ \mathbf{y} \end{bmatrix} - 2 \begin{bmatrix} \mathbf{0} \\ w_{r+1}^{(r)} \\ \mathbf{u} \end{bmatrix} \{\bar{w}_{r+1}^{(r)} x_{r+1} + \mathbf{u}^*\mathbf{y}\} = \begin{bmatrix} \hat{\mathbf{x}} \\ x_{r+1} - 2zw_{r+1}^{(r)} \\ \mathbf{y} - 2z\mathbf{u} \end{bmatrix}$$

where

$$z = (\bar{w}_{r+1}^{(r)} x_{r+1} + \mathbf{u}^*\mathbf{y}).$$

Obviously in order that the $r+2$ to n elements of $P^{(r)}\mathbf{x}$ are zero we must choose $\mathbf{u} = \alpha\mathbf{y}$, where α is constant, and set

$$1-2\alpha z = 0. \tag{5}$$

We also require $\{\mathbf{w}^{(r)}\}^* \mathbf{w}^{(r)} = 1$, giving

$$|\alpha|^2\mathbf{y}^*\mathbf{y}+|w_{r+1}^{(r)}|^2 = 1. \tag{6}$$

Since the vector $\mathbf{x}$ is real we assume α, $\mathbf{w}^{(r)}$ are real and set $w_{r+1}^{(r)} = \alpha v_{r+1}^{(r)}$. From equations (5) and (6) it follows that α and $v_{r+1}^{(r)}$ satisfy

$$1-\alpha^2[2v_{r+1}^{(r)}x_{r+1}+2\mathbf{y}^*\mathbf{y}] = 0$$

$$1-\alpha^2[\{v_{r+1}^{(r)}\}^2+\mathbf{y}^*\mathbf{y}] = 0.$$

Elimination of α gives

$$v_{r+1}^{(r)} = x_{r+1}\pm\{(x_{r+1})^2+\mathbf{y}^*\mathbf{y}\}^{1/2}$$

$$= x_{r+1}\pm s, \text{ say,}$$

the sign being chosen to agree with that of x_{r+1}, in order to reduce cancellation. The factor α is then given by

$$\alpha = 2^{-1/2}\{s^2\pm x_{r+1}s\}^{-1/2}$$

from which $w_{r+1}^{(r)}$ may be obtained. The process is easily arranged (see Section 3.4) so that only one square root need be calculated for every transformation. In addition, the number of multiplications required to complete the tridiagonalization is half that required in Givens' process.

Example *Apply Householder's method to tridiagonalize the matrix*

$$A = \begin{bmatrix} 1 & 2 & 1 & 2 \\ 2 & 2 & -1 & 1 \\ 1 & -1 & 1 & 1 \\ 2 & 1 & 1 & 1 \end{bmatrix}$$

$r = 1$ Here we have $\mathbf{x}^T = (1, 2, 1, 2)$ and so $x_2 = 2$, $\mathbf{y}^T = (1, 2)$. Thus $s = 3$, $v_2^{(1)} = 5$ and $\alpha^2 = \frac{1}{30}$, giving $\mathbf{w}^{(1)} = \alpha(0, 5, 1, 2)^T$. Using exact arithmetic, we have

$$A_2 = \begin{bmatrix} 1 & -3 & 0 & 0 \\ -3 & \frac{7}{3} & \frac{7}{15} & -\frac{1}{15} \\ 0 & \frac{7}{15} & \frac{118}{75} & \frac{101}{75} \\ 0 & -\frac{1}{15} & \frac{101}{75} & \frac{7}{75} \end{bmatrix}.$$

$r = 2$ Here

$$\mathbf{x}^T = (-3, \tfrac{7}{3}, \tfrac{7}{15}, -\tfrac{1}{15})$$

and so

$$\mathbf{x}_3 = \tfrac{7}{15}, \mathbf{y}^T = (-\tfrac{1}{15}).$$

Thus

$$s = \frac{\sqrt{50}}{15} = 0{\cdot}4714, \; v_3^{(2)} = 0{\cdot}9381 \text{ and } \alpha^2 = 1{\cdot}13067,$$

(rounding to the number of figures shown), giving

$$\mathbf{w}^{(2)} = \alpha(0, 0, 0{\cdot}9381, -0{\cdot}0667)^T.$$

Thus, to four decimal place accuracy where necessary, we have finally

$$A_3 = \begin{bmatrix} 1 & -3 & 0 & 0 \\ -3 & 2{\cdot}3333 & -0{\cdot}4714 & 0 \\ 0 & -0{\cdot}4714 & 1{\cdot}1667 & -1{\cdot}5000 \\ 0 & 0 & -1{\cdot}5000 & 0{\cdot}5000 \end{bmatrix}$$

Note that this matrix differs from the tridiagonal matrix obtained by applying Givens' method to the extent of a similarity transformation with the diagonal matrix

$$\begin{bmatrix} -1 & & & \\ & 1 & & \\ & & 1 & \\ & & & -1 \end{bmatrix}.$$

Despite the greater efficiency of the Householder reduction process for a dense matrix, there are occasions where the Givens process may be more efficient. It is relatively easy in the latter method to ignore a transformation if the element $a_{jk}^{(r)}$ is already zero. In Householder's method, little advantage is gained by testing for zero entries. The Givens strategy is sometimes important in the reduction of sparse matrices to tridiagonal form.

It should be noted that the transformations in each reduction need only be stored if the eigenvectors of the matrix A are required. In such a situation, it is necessary to back-transform the eigenvectors of the tridiagonal matrix to obtain the eigenvectors of A.

8.4 Reduction of a Hermitian matrix

The above two techniques may be readily extended to cover the reduction of a Hermitian matrix to either Hermitian tridiagonal form or to real symmetric tridiagonal form. Since the Householder process is more efficient we restrict our consideration to its generalization. In fact we may carry out the reduction in one of two ways. If we restrict ourselves to the use of elementary Hermitian matrices then in general we can only reduce A to Hermitian tridiagonal form (see exercise 12, Section 3.4). However a further similarity transformation with a unitary diagonal matrix will convert it into real symmetric form (exercise 9, Section 2.3). A more pleasing approach is to generalize the notion of an elementary Hermitian matrix.

Let

$$P_\lambda = I - \lambda \mathbf{w}\mathbf{w}^*$$

where $\mathbf{w}^*\mathbf{w} = 1$ and λ is a complex constant. The matrix P_λ is unitary if $\lambda + \bar{\lambda} = \lambda\bar{\lambda}$ but only Hermitian if $\lambda = 2$ (see Section 3.4). We have shown that it is possible to choose $\mathbf{w}$ and λ so that for an arbitrary complex vector $\mathbf{x}$ we have

$$P_\lambda \mathbf{x} = k\mathbf{e}_1 \quad \text{where } k \text{ is real.}$$

The analysis proceeds as outlined in Section 3.4. Since P_λ is unitary it follows that

$$\bar{k}k = \mathbf{x}^*\mathbf{x}$$

and as k is real we choose

$$k = \{\mathbf{x}^*\mathbf{x}\}^{1/2}.$$

Now from Section 3.4,

$$\mathbf{w} = \frac{1}{\alpha}(\mathbf{x} - k\mathbf{e}_1)$$

where

$$\begin{aligned} \alpha\bar{\alpha} &= \mathbf{x}^*\mathbf{x} - k\mathbf{e}_1^T\mathbf{x} - k\mathbf{x}^*\mathbf{e}_1 + k^2 \\ &= 2k^2 - 2ku \end{aligned}$$

where $x_1 = u + iv$, $i = \sqrt{-1}$. Moreover

$$\lambda = \frac{\alpha\bar{\alpha}}{(\mathbf{x} - k\mathbf{e}_1)^*\,\mathbf{x}} = \frac{2(k-u)}{k - x_1}$$

and it follows also that

$$\begin{aligned} w_1 &= (x_1 - k)/\alpha, \\ w_i &= x_i/\alpha, \quad i = 2, \ldots, n. \end{aligned}$$

Using this result it is now a simple matter to carry out the reduction of a Hermitian matrix to real symmetric tridiagonal form by $(n-1)$ transformations of the form P_λ.

Exercises

4. Show that it is possible to choose a sequence of unitary matrices P_λ so as to decompose an arbitrary complex matrix into the product of a unitary matrix and an upper triangular matrix with real positive diagonal elements.

5. Carry out the reduction, by Householder's method, of the matrix given in exercise 2 of Section 8.2, and compare the results.

9

Eigensystem of a symmetric tridiagonal matrix

9.1 Introduction

The Givens and Householder transformations are used to reduce a symmetric (or Hermitian) matrix to symmetric tridiagonal form. There remains the problem of determining the eigensystem of this simpler form. At this stage we must decide whether we require the complete eigensystem (that is all eigenvalues with or without their corresponding eigenvectors) or simply information on a few eigenvalues and their related eigenvectors. We include with the latter the problem of finding information on the distribution or clustering of eigenvalues. A most efficient algorithm for determining the complete eigensystem will be discussed in Chapter 10. In this chapter, we study the second type of problem. Possible questions that we could ask include the following:

(i) What is the eigenvalue nearest to a particular point on the real line? (This problem has already been considered in Chapter 6, and is probably best treated by inverse iteration.)

(ii) How many eigenvalues of the matrix lie in a particular interval of the real axis?

(iii) Are there any clusters of eigenvalues and how many eigenvalues are there in each cluster?

(iv) Are there any very small (very large) eigenvalues?

It can be seen that the questions we are asking in this chapter are more specific than those we have considered before, or will consider later. We may not now be particularly interested in the *value* of the eigenvalue but rather in its *relative position* in the spectrum of eigenvalues. However the technique to be developed may readily be used to compute an approximation to any particular eigenvalue.

Since we are dealing with real symmetric tridiagonal matrices we may simplify our notation somewhat. Let T denote a symmetric tridiagonal matrix with elements

$$t_{ii} = d_i$$

$$t_{i,i+1} = t_{i+1,i} = e_i.$$

We may assume that none of the e_i is zero, for otherwise we could subdivide T into the direct sum of tridiagonal matrices of lower order and work instead with them.

Thus

$$T = \begin{bmatrix} d_1 & e_1 & & & & \\ e_1 & d_2 & e_2 & & & \\ & e_2 & d_3 & e_3 & & \\ & & \cdot & \cdot & \cdot & \\ & & & \cdot & \cdot & \cdot \\ & & & & \cdot & \cdot & \cdot \\ & & & & e_{n-2} & d_{n-1} & e_{n-1} \\ & & & & & e_{n-1} & d_n \end{bmatrix} = [e_{i-1}, d_i, e_i]_1^n,$$

the notation $[a_{i-1}, b_i, c_i]_1^n$ frequently being employed to denote a tridiagonal matrix.

9.2 Sturm sequences and bisection

Let $p_r(\lambda)$ denote the determinant of the leading principal minor of $T-\lambda I$. Thus

$$p_r(\lambda) = \det \begin{bmatrix} d_1-\lambda & e_1 & & & & \\ e_1 & d_2-\lambda & e_2 & & & \\ & \cdot & \cdot & \cdot & & \\ & & \cdot & \cdot & \cdot & \\ & & & \cdot & \cdot & \cdot \\ & & & e_{r-2} & d_{r-1}-\lambda & e_{r-1} \\ & & & & e_{r-1} & d_r-\lambda \end{bmatrix}.$$

We define $p_0(\lambda) = 1$ and by inspection observe that

$$p_1(\lambda) = d_1-\lambda.$$

Expanding $p_r(\lambda)$ by the final row the result

$$p_r(\lambda) = (d_r-\lambda)\, p_{r-1}(\lambda) - e_{r-1}^2 p_{r-2}(\lambda)$$

follows by simple determinantal relations. Since

$$p_n(\lambda) = \det(T-\lambda I)$$

we may compute it by means of the relations

$$p_0(\lambda) = 1$$

$$p_1(\lambda) = d_1 - \lambda \tag{1}$$

$$p_r(\lambda) = (d_r - \lambda)\, p_{r-1}(\lambda) - e_{r-1}^2 p_{r-2}(\lambda), \quad r = 2, 3, \ldots, n.$$

Now it may be shown that the zeros of $p_r(\lambda)$ strictly separate those of $p_{r-1}(\lambda)$, provided none of the e_i is zero (see exercise 4). This property forms the basis of a powerful technique for determining the positions of the eigenvalues of T. We evaluate, by means of the formulas (1), the numbers $p_0(\lambda), \ldots, p_n(\lambda)$ for some value of λ. We denote by $s(\lambda)$ the number of agreements in sign between successive members of the sequence $\{p_r(\lambda)\}$, e.g. $++-+$ gives $s = 1$. (If $p_r(\lambda)$ is zero then its sign is taken to be opposite to that of $p_{r-1}(\lambda)$.)

Then the main result may be stated as follows:

Theorem 9.1 (*Sturm sequence property*) *The number of agreements in sign $s(\lambda)$ of successive members of the sequence $\{p_r(\lambda)\}$ is equal to the number of eigenvalues of T which are strictly greater than λ.*

The proof of this theorem is relatively easy and is left as exercise 6.

We note in passing that a more general result can in fact be proved. This is that the property is possessed by the sequence of determinants of the leading principal minors of the matrix $C - \lambda I$, where C is *any* symmetric matrix. We will use this result in Chapter 14.

A systematic search using Theorem 9.1 for several values of λ allows us to locate approximate intervals in which each of the eigenvalues lie. Thus if we find that

$$s(\lambda_1) = k_1, \qquad s(\lambda_2) = k_1 + 1 \qquad (\lambda_2 < \lambda_1)$$

then it follows that there is only one eigenvalue in the interval $[\lambda_2, \lambda_1]$.

Without any further refinement the Sturm sequence property allows us to answer questions such as the following:

(i) Is the matrix positive (negative) definite?
(ii) How many eigenvalues of the matrix lie in the interval $[a, b]$?
(iii) What is a rough upper bound on the spectral radius of the matrix?
(iv) Is the matrix singular (nearly singular)?

Exercise

1. Indicate how the above questions may be answered.

Example *Is the matrix*

$$T = \begin{bmatrix} -2 & 1 & & \\ 1 & -2 & 1 & \\ & 1 & -2 & 1 \\ & & 1 & -2 \end{bmatrix}$$

negative definite? How many eigenvalues lie in the interval $[-2, 0]$?

Here $d_i = -2$, $e_i = 1$.
Let $\lambda = 0$

$$\begin{aligned} p_0(0) &= 1 \\ p_1(0) &= -2 \\ p_2(0) &= (-2)(-2)-1 = 3 \\ p_3(0) &= (-2)3-1(-2) = -4 \\ p_4(0) &= (-2)(-4)-1(3) = 5. \end{aligned}$$

The signs of the sequence $\{p_r(0)\}$ alternate and $s(0) = 0$. It follows that T is negative definite. Notice that Gerschgorin's theorem (Theorem 2.4) only tells us that T is negative semidefinite. Using $\lambda = -2$ gives the sequence

$$\begin{aligned} p_0(-2) &= 1 \\ p_1(-2) &= -2+2 = 0 \\ p_2(-2) &= 0-1 = -1 \\ p_3(-2) &= 0-0 = 0 \\ p_4(-2) &= 0+1 = 1 \end{aligned}$$

This time we have two $p_r(\lambda)$ with zero value. Applying the rule given above, the sequence of signs is determined as

$$+ - - + + .$$

We have two agreements in sign and it follows that T has two eigenvalues in $[-2, 0]$.

Using the Sturm sequence property together with the *method of bisection* allows us to determine any eigenvalue to prescribed accuracy. The first step is to locate an interval $[\lambda_1^{(0)}, \lambda_2^{(0)}]$ in which only the eigenvalue λ lies, by means of the Sturm sequence procedure. It follows that for some $k<n$ we have that

$$s(\lambda_2^{(0)}) = k, \qquad s(\lambda_1^{(0)}) = k+1.$$

Let $\mu^{(0)} = \frac{1}{2}[\lambda_1^{(0)}+\lambda_2^{(0)}]$ and evaluate $s(\mu^{(0)})$ which must either have the value k or $k+1$. If $s(\mu^{(0)}) = k$, then λ lies in the interval $[\lambda_1^{(0)}, \mu^{(0)}]$; otherwise it is in $[\mu^{(0)}, \lambda_2^{(0)}]$. This process may be repetitively applied to determine a tight interval for λ. The procedure may be stated as follows for a general step based on the interval $[\lambda_1^{(j)}, \lambda_2^{(j)}]$:

(i) Define $\mu^{(j)} = \frac{1}{2}[\lambda_1^{(j)}+\lambda_2^{(j)}]$.
(ii) Evaluate $s(\mu^{(j)})$.
(iii) If $s(\mu^{(j)}) = s(\lambda_1^{(j)})$ then set

$$\lambda_1^{(j+1)} = \mu^{(j)}, \quad \lambda_2^{(j+1)} = \lambda_2^{(j)}$$

otherwise set

$$\lambda_1^{(j+1)} = \lambda_1^{(j)}, \quad \lambda_2^{(j+1)} = \mu^{(j)}.$$

(iv) Test to see if $|\lambda_1^{(j+1)}-\lambda_2^{(j+1)}| < \epsilon$ (required accuracy). If the test fails return to step (i).

We note that

$$|\lambda_1^{(j)}-\lambda_2^{(j)}| = 2^{-j}|\lambda_1^{(0)}-\lambda_2^{(0)}|,$$

so in order to satisfy the test in (iv) we require

$$j > \frac{\ln\left[\dfrac{|\lambda_1^{(0)}-\lambda_2^{(0)}|}{\epsilon}\right]}{\ln 2}.$$

The above technique is very efficient if we require to find the eigenvalues in a particular interval, or the first few, or last few eigenvalues of an $n \times n$ symmetric tridiagonal matrix.

Exercises

2. Continue the analysis of the example in the text to find intervals in which each of the eigenvalues of the 4×4 tridiagonal matrix lie.

3. How would you interpret the occurrence $p_n(\lambda) = 0$ for some λ for an $n \times n$ tridiagonal matrix?

4. *Separation theorem* Prove that the zeros of $p_r(\lambda)$ strictly separate those of $p_{r+1}(\lambda)$ if no off-diagonal element is zero. (Hint: first of all show that the zero of $p_1(\lambda)$ separates those of $p_2(\lambda)$ and then use induction on equations (1).)

5. Show that if all the e_i are non-zero, the eigenvalues are distinct.

6. Prove Theorem 9.1. (Hint: clearly $s(-\infty) = n$; assuming all the e_i are non-zero, increase λ away from $-\infty$, and use exercises 4 and 5.)

7. If we let $b_r(\lambda)$ denote the determinant of the bottom $(n-r+1)\times(n-r+1)$ submatrix of the matrix $T-\lambda I$ show that the following relations hold

$$b_0(\lambda) = 1$$

$$b_1(\lambda) = d_n-\lambda$$

$$b_r(\lambda) = (d_{n-r+1}-\lambda)\, b_{r-1}(\lambda)-e_{n-r+1}^2 b_{r-2}(\lambda).$$

Do these functions also have the Sturm sequence property?

8. Given the tridiagonal matrix

$$\begin{bmatrix} 1 & 2 & 0 \\ 2 & -1 & 1 \\ 0 & 1 & 3 \end{bmatrix}$$

use the Sturm sequence bisection method to find all the eigenvalues correct to two decimal places.

9.3 Eigenvectors of a tridiagonal matrix

The computation of the eigenvectors associated with the eigenvalues computed by means of the technique in the previous section should be undertaken by inverse iteration as outlined in Chapter 6. The decomposition of the matrix $T-\lambda I$ can be efficiently determined since the matrix is tridiagonal.

However it is sometimes tempting to employ a simpler technique which unfortunately may lead to disastrous consequences. The eigenvector $\mathbf{x}$ corresponding to the eigenvalue λ is given by the set of equations

$$\begin{bmatrix} d_1-\lambda & e_1 & & & \\ e_1 & d_2-\lambda & e_2 & & \\ & \cdot & \cdot & & \\ & & \cdot & \cdot & \\ & & \cdot & \cdot & \\ & & & d_{n-1}-\lambda & e_{n-1} \\ & & & e_{n-1} & d_n-\lambda \end{bmatrix} \begin{bmatrix} x_1 \\ x_2 \\ \cdot \\ \cdot \\ \cdot \\ \\ x_n \end{bmatrix} = 0$$

where λ is known to some precision. Since $\mathbf{x}$ is arbitrary to a scalar multiple it would appear possible to set one of the components x_i to a preassigned number (of course this technique would fail if we set to a non-zero quantity an element which was actually zero in the eigenvector). For example we could choose $x_n = 1$ as this cannot be zero. Then the above set of equations would recursively determine $x_{n-1}, x_{n-2}, \ldots, x_1$. Alternatively we may set $x_1 = 1$ (which again cannot be zero) and determine the remaining $x_2, x_3, \ldots, x_n$. However this process is in general dangerously unstable. Wilkinson (1965) gives a detailed analysis of these recursions but we shall content ourselves with a simple warning. *Under no circumstances should one use the backward (or forward) recursion process in preference to inverse iteration.*

In this chapter, we have introduced a technique for the partial solution to the symmetric tridiagonal eigenproblem. In the following chapter, we consider one of the most powerful techniques yet devised for the solution of the complete tridiagonal eigenproblem.

10

The LR and QR algorithms

10.1 Introduction

In the preceding chapter, an algorithm employing Sturm sequences and the technique of bisection was introduced to determine the eigenvalues of a symmetric tridiagonal matrix. In this chapter, we turn to another type of algorithm which may be employed to solve the same problem. We will assume throughout this chapter that the original Hermitian matrix has been reduced to symmetric tridiagonal form by one of the techniques outlined in Chapter 8.

Although the introduction of the LR algorithm was probably one of the landmarks in the development of powerful algorithms we will not discuss it in detail. Its introduction led to the much more important QR algorithm which has generally been accepted as the most powerful technique for the solution of the symmetric eigenproblem.

10.2 The LR algorithm

This algorithm, developed by H. Rutishauser in 1958, is based on the LU decomposition of a matrix. (We now use R in place of U to denote an upper triangular matrix.) Suppose we construct (if possible) for the matrix A_1 the decomposition

$$A_1 = L_1R_1$$

where L_1 is unit lower triangular, and then form the matrix A_2 by reversing the order of the matrix product, thus

$$A_2 = R_1L_1.$$

By substitution it follows that

$$A_2 = L_1^{-1}A_1L_1$$

so that the matrix A_2 is similar to the matrix A_1. Applying this strategy iteratively gives rise to the algorithm

$$A_s = L_sR_s, \qquad A_{s+1} = R_sL_s,$$

which is the fundamental LR algorithm. Its importance lies in the fact that under certain conditions the sequence of matrices A_s tends to an upper triangular matrix whose diagonal elements converge to the eigenvalues of the matrix A_1. The necessary conditions and proof of convergence may be found in the book of Wilkinson (1965).

Exercise

1. By using the result

$$A_{s+1} = L_s^{-1} A_s L_s$$

show that

$$A_1^s = L^{(s)} R^{(s)}$$

where

$$L^{(s)} = L_1 L_2 \dots L_s, \quad R^{(s)} = R_s R_{s-1} \dots R_1.$$

(Hint: prove that $L^{(s)} A_s = A_1 L^{(s)}$.)

It is easy to see the major shortcoming of the LR process. As we pointed out in Section 3.5 it is sometimes essential and always desirable to include partial pivoting in the algorithm. Thus the LR algorithm as so far defined does not always exist. A generalization of the algorithm which includes interchanges is easily definable. Unfortunately this generalization is still not sufficient to ensure that a modified LR algorithm is definable for all matrices A. (See exercise 3 of this section.)

An extension of the LR algorithm which produces more rapid convergence but which, however, may still not exist, is given by

$$A_s - k_s I = L_s R_s$$

$$A_{s+1} = k_s I + R_s L_s$$

where k_s is a suitably chosen constant. Thus we effectively introduce a shift of origin. It may easily be verified that the sequence A_s is similar to A_1 in the sense that

$$A_{s+1} = L_s^{-1} A_s L_s.$$

Corresponding to exercise 1 above we may verify that

$$L^{(s)} R^{(s)} = \prod_{i=1}^{s} (A_i - k_i I).$$

Exercises

2. The above LR process with 'shifts' is known as a *restoring process*. In computations the following non-restoring process may be employed:

$$A_s - k_s I = L_s R_s$$

$$A_{s+1} = R_s L_s.$$

Prove that

$$L^{(s)}R^{(s)} = A_1 - \sum_{i=1}^{s} k_i I.$$

3. Verify that the LR decomposition does not exist for matrices of the form

$$\begin{bmatrix} \lambda & b \\ a & \lambda \end{bmatrix}$$

for some value of λ.

The choice of the shifts k_s must be made to ensure rapid convergence. We leave discussion of choices of shifts until a later section. In fact we conclude our discussion of the LR algorithm at this stage and turn instead to the much more important analogue based on the unitary decomposition of a matrix, developed by J. G. F. Francis in 1961.

10.3 The QR algorithm

In Chapter 3, it was shown how an arbitrary matrix A could be decomposed into the product of a unitary matrix Q and an upper triangular matrix R. This decomposition was achieved by means of plane rotations, or by means of elementary Hermitian matrices.

The main deficiency of the LR procedure is the possibility of breakdown in the LU decomposition. There is no danger of breakdown in the QR decomposition of a matrix. This conclusion is immediate from consideration of the process whereby the decomposition is constructed.

The unshifted algorithm is defined by the relations

$$A_s = Q_s R_s, \qquad A_{s+1} = R_s Q_s.$$

Of course in practice we actually determine the unitary matrix Q_s^* so that

$$Q_s^* A_s = R_s.$$

The sequence of matrices A_s are unitarily similar to each other and hence to $A_1 \equiv A$ since

$$A_{s+1} = R_s Q_s = Q_s^* A_s Q_s.$$

Further we have that

$$\begin{aligned} A_{s+1} = Q_s^* A_s Q_s &= Q_s^* Q_{s-1}^* A_{s-1} Q_{s-1} Q_s \\ &= (Q_s^* \ldots Q_1^*)\, A_1 (Q_1 \ldots Q_s) \end{aligned}$$

or

$$Q_1 \ldots Q_s A_{s+1} = A_1 Q_1 \ldots Q_s.$$

Defining

$$Q^{(s)} = Q_1 \ldots Q_s,$$

$$R^{(s)} = R_s \ldots R_1,$$

we have that

$$\begin{aligned} Q^{(s)}R^{(s)} &= Q_1 \dots Q_{s-1}Q_sR_sR_{s-1} \dots R_1 \\ &= Q_1 \dots Q_{s-1}A_sR_{s-1} \dots R_1 \\ &= A_1Q_1 \dots Q_{s-1}R_{s-1} \dots R_1 \\ &= A_1Q^{(s-1)}R^{(s-1)} \end{aligned}$$

and finally that

$$Q^{(s)}R^{(s)} = A_1^s.$$

It follows that $Q^{(s)}$ and $R^{(s)}$ give the unitary decomposition of A_1^s, and moreover this decomposition is unique if we insist on the diagonal elements of the upper triangular matrices R_j being positive.

A very important property of the QR algorithm is that all the matrices A_s are of tridiagonal form if A_1 is of tridiagonal form. (Thus we see that the reduction of a Hermitian matrix to a tridiagonal form leads to an efficient QR algorithm.) To verify this invariance in the structure of the A_s let us consider a 5×5 tridiagonal matrix

$$A_1 = \begin{bmatrix} x & x & & & \\ x & x & x & & \\ & x & x & x & \\ & & x & x & x \\ & & & x & x \end{bmatrix}.$$

The matrix A_1 is decomposed into the form Q_1R_1 by means of rotations in the planes (1, 2), (2, 3), (3, 4), (4, 5) to annihilate the elements in the positions (2, 1), (3, 2), (4, 3), (5, 4) respectively.
It follows that the matrix R_1 has the form

$$R_1 = \begin{bmatrix} x & x & x & & \\ & x & x & x & \\ & & x & x & x \\ & & & x & x \\ & & & & x \end{bmatrix}.$$

To complete the similarity transformation we postmultiply R_1 by Q_1. This corresponds to rotations (columnwise) in planes (1, 2), (2, 3), (3, 4), (4, 5).

Thus the matrix $A_2 = R_1Q_1$ must have the form

$$A_2 = \begin{bmatrix} x & x & x & & \\ x & x & x & x & \\ & x & x & x & x \\ & & x & x & x \\ & & & x & x \end{bmatrix}.$$

However the two steps of the QR algorithm correspond to a similarity transformation of the symmetric matrix A_1 and therefore A_2 must be symmetric. It follows immediately that A_2 must be symmetric tridiagonal and hence that the QR algorithm preserves the structure of A_1.

Exercise

4. Show that in general if A_1 is a $(2m+1)$ band symmetric matrix then in the QR process all A_s are $(2m+1)$ band symmetric matrices.

10.4 The QR algorithm with shifts

The rate of convergence of the unshifted algorithm is in general not sufficiently fast to make it useful in practice. However as mentioned in connection with the LR algorithm it is standard to incorporate shifts which produce more rapid convergence.

The shifted QR algorithm with restoring is given by

$$\begin{aligned} A_s - k_sI &= Q_sR_s \\ A_{s+1} &= k_sI + R_sQ_s \end{aligned} \tag{1}$$

for a suitably chosen real number k_s. Since we have assumed that the matrix A_1 is symmetric tridiagonal all the matrices A_s are symmetric tridiagonal and we may simplify our notation by writing

$$a_{kk}^{(s)} = d_k^{(s)},\ a_{k,k+1}^{(s)} = a_{k+1,k}^{(s)} = e_{k+1}^{(s)}$$

where $a_{ij}^{(s)}$ is the (i, j) element of A_s. Further it was shown in exercise 5 of Chapter 9 that if all the $e_{k+1}^{(1)}$ are non-zero then the matrix A_1 has distinct eigenvalues. We shall assume that this is the case for otherwise the matrix A_1 would break into several tridiagonal matrices of smaller size which we could treat independently. So far the convergence of the QR process has been discussed without reference to what we precisely mean by convergence. If the shifts k_s are chosen according to one of two rules then the sequence of matrices A_s converges to a matrix which is block diagonal in form, the blocks being either one by one or two by two. This latter case corresponds to eigenvalues of equal modulus but opposite sign.

There are two common choices of origin shift. The first consists of choosing $k_s = d_n^{(s)}$, the element in the bottom right-hand corner of the matrix A_s. With this choice, whilst convergence is guaranteed, and is ultimately cubic, it is generally slow. Since the second choice of origin shift produces faster general convergence, and is generally preferred in practice we shall study it in greater detail. This second choice of shift corresponds to choosing $k_s = \lambda_s$ where λ_s is the eigenvalue of the two by two matrix

$$\begin{bmatrix} d_{n-1}^{(s)} & e_n^{(s)} \\ e_n^{(s)} & d_n^{(s)} \end{bmatrix}$$

which is closer to $d_n^{(s)}$. The proof of convergence which we give in the next section follows closely that given in the paper by Wilkinson (1968). This paper also gives a proof of convergence for the first shift strategy.

For either of the above shifts we find that

$$e_n^{(s)} e_{n-1}^{(s)} \to 0$$

as $s \to \infty$. If $e_n^{(s)}$ becomes negligible it follows that we may take $d_n^{(s)}$ to be an eigenvalue and then omit the final row and column of the matrix A_s. If $e_{n-1}^{(s)}$ becomes negligible while $e_n^{(s)}$ is significant we may determine two eigenvalues from the two by two matrix in the bottom right-hand corner of A_s, omit the final two rows and columns of A_s and proceed. We see that the algorithm produces a deflation in the order of the matrices. This leads to extremely rapid convergence rates.

10.5 Analysis of convergence†

We give a proof of convergence for the QR algorithm using the second shift strategy. A basic result is contained in Lemma 10.1.

Lemma 10.1 *For the QR process* (1), *all elements* $d_k^{(s)}$ *and* $e_k^{(s)}$ *are bounded by* $\max|\lambda_i|$ *and all elements derived during the triangularization of all* A_s *including all elements of* R_s *are bounded by* $2\max|\lambda_i|$ *where* λ_i *is an eigenvalue of* A_1.

Proof The A_s are unitarily (in fact orthogonally since A_s, Q_s, R_s are all real) similar to A_1. The spectral radius of a matrix is invariant under a similarity transformation and thus

$$\|A_s\|_2 = \|A_1\|_2 = \max|\lambda_i|.$$

But by definition every element of A_s is bounded by $\|A_s\|_2$ and thus

$$|d_k^{(s)}| \leqslant \max|\lambda_i|$$
$$|e_k^{(s)}| \leqslant \max|\lambda_i|$$

for all k, s.

† This section may be omitted on a first reading.

Since the shift k_s is an eigenvalue of a principal submatrix of A_s it follows from the separation theorem (see Section 9.2) that

$$|k_s| \leqslant \max|\lambda_i|.$$

Thus, using the triangle inequality

$$||A_s - k_s I||_2 \leqslant ||A_s||_2 + ||k_s I||_2 \leqslant 2\max|\lambda_i|.$$

Now the Euclidean norm of any column or row of a matrix is bounded by the spectral norm of the matrix, and also in the unitary decomposition of $A_s - k_s I$ to R_s the Euclidean norm of each column is invariant. Thus all elements derived during all triangularizations are bounded by 2 max$|\lambda_i|$. The lemma is thus proved.

The tridiagonal matrix $A_s - k_s I$ is reduced to upper triangular form by premultiplication by plane rotations. We have in fact that

$$R^{(s)}(n-1, n)\, R^{(s)}(n-2, n-1) \ldots R^{(s)}(2, 3)\, R^{(s)}(1, 2)(A_s - k_s I) = R_s$$

where $R^{(s)}(i, j)$ is a rotation in the (i, j) plane. Let $c_{i+1}^{(s)}, s_{i+1}^{(s)}$ denote the cosine and sine associated with $R^{(s)}(i, i+1)$. Also define $\hat{d}_i^{(s)} = d_i^{(s)} - k_s$. Immediately prior to premultiplication by $R^{(s)}(k-1, k)$ the configuration is given by

$$\begin{bmatrix} p_1^{(s)} & q_1^{(s)} & r_1^{(s)} & & & & & & & \\ & p_2^{(s)} & q_2^{(s)} & r_2^{(s)} & & & & & & \\ & & \cdot & \cdot & \cdot & & & & & \\ & & & \cdot & \cdot & \cdot & & & & \\ & & & & \cdot & \cdot & \cdot & & & \\ & & & & p_{k-2}^{(s)} & q_{k-2}^{(s)} & r_{k-2}^{(s)} & & & \\ & & & & & x_{k-1}^{(s)} & y_{k-1}^{(s)} & & & \\ & & & & & e_k^{(s)} & \hat{d}_k^{(s)} & e_{k+1}^{(s)} & & \\ & & & & & & \cdot & \cdot & \cdot & \\ & & & & & & & \cdot & \cdot & \cdot \\ & & & & & & & & \cdot & \cdot \quad \cdot \\ & & & & & & & e_{n-1}^{(s)} & \hat{d}_{n-1}^{(s)} & e_n^{(s)} \\ & & & & & & & & e_n^{(s)} & \hat{d}_n^{(s)} \end{bmatrix}.$$

Carrying out the transformation gives

$$\begin{aligned} p_{k-1}^{(s)} &= \{(x_{k-1}^{(s)})^2 + (e_k^{(s)})^2\}^{1/2} \\ c_k^{(s)} &= \frac{x_{k-1}^{(s)}}{p_{k-1}^{(s)}},\; s_k^{(s)} = \frac{e_k^{(s)}}{p_{k-1}^{(s)}} \\ x_k^{(s)} &= \hat{d}_k^{(s)} c_k^{(s)} - y_{k-1}^{(s)} s_k^{(s)} \\ y_k^{(s)} &= e_{k+1}^{(s)} c_k^{(s)}. \end{aligned} \tag{2}$$

Successive reduction leads to the matrix

$$R_s = \begin{bmatrix} p_1^{(s)} & q_1^{(s)} & r_1^{(s)} & & & \\ & \cdot & \cdot & \cdot & & \\ & & \cdot & \cdot & \cdot & \\ & & & \cdot & \cdot & \cdot \\ & & & & p_{n-1}^{(s)} & q_{n-1}^{(s)} \\ & & & & & x_n^{(s)} \end{bmatrix}.$$

One step of the QR algorithm requires the postmultiplication of R_s by $(R^{(s)}(1, 2))^T\,(R^{(s)}(2, 3))^T \ldots (R^{(s)}(n-1, n))^T$. This gives rise to the relations

$$\begin{aligned} e_i^{(s+1)} &= p_i^{(s)}s_i^{(s)} \quad i = 2, 3, \ldots, n-1 \\ e_n^{(s+1)} &= x_n^{(s)}s_n^{(s)} \\ d_n^{(s+1)} &= k_s + x_n^{(s)}c_n^{(s)} \end{aligned} \tag{3}$$

which are fundamental to our analysis. From the definition of our shift strategy we have

$$\hat{d}_n^{(s)}\hat{d}_{n-1}^{(s)} = \{e_n^{(s)}\}^2$$

and

$$|d_n^{(s)}| \leqslant |d_{n-1}^{(s)}|$$

from which it follows that

$$|d_{n-1}^{(s)}| \geqslant |e_n^{(s)}|.$$

In order to prove convergence, we must show that $|e_{n-1}^{(s)}e_n^{(s)}| \to 0$. We first of all prove that $|e_{n-1}^{(s)}e_n^{(s)}|$ is monotonic decreasing. From the relations (2) and (3) we have that

$$\begin{aligned} x_n^{(s)} &= \left(\frac{e_n^{(s)}}{\hat{d}_{n-1}^{(s)}}\right)(-s_n^{(s)}y_{n-2}^{(s)}s_{n-1}^{(s)}) \\ e_n^{(s+1)} &= x_n^{(s)}s_n^{(s)} \end{aligned}$$

with

$$y_{n-2}^{(s)} = e_{n-1}^{(s)}c_{n-2}^{(s)}.$$

It follows that

$$e_n^{(s+1)} = \left(\frac{e_n^{(s)}}{\hat{d}_{n-1}^{(s)}}\right)(-\{s_n^{(s)}\}^2\, e_{n-1}^{(s)}c_{n-2}^{(s)}s_{n-1}^{(s)}).$$

Since the quantities $e_n^{(s)}/\hat{d}_{n-1}^{(s)}$, $s_n^{(s)}$, $c_{n-2}^{(s)}$, $s_{n-1}^{(s)}$ are all bounded in modulus by unity we have that

$$|e_n^{(s+1)}| \leqslant |e_{n-1}^{(s)}|.$$

Also since

$$e_{n-1}^{(s+1)} = p_{n-1}^{(s)} s_{n-1}^{(s)}$$

we have that

$$e_{n-1}^{(s+1)} e_n^{(s+1)} = e_{n-1}^{(s)} e_n^{(s)} \left\{ \frac{-p_{n-1}^{(s)} c_{n-2}^{(s)} (s_n^{(s)} s_{n-1}^{(s)})^2}{\hat{d}_{n-1}^{(s)}} \right\}.$$

If the substitution $p_{n-1}^{(s)} = e_n^{(s)}/s_n^{(s)}$ is now made the result

$$|e_{n-1}^{(s+1)} e_n^{(s+1)}| = |e_{n-1}^{(s)} e_n^{(s)}| \, |c_{n-2}^{(s)} s_n^{(s)} \{s_{n-1}^{(s)}\}^2| \, (|e_n^{(s)}|/|\hat{d}_{n-1}^{(s)}|) \tag{4}$$

is obtained. The final two quantities on the right-hand side are bounded by unity and thus

$$|e_{n-1}^{(s+1)} e_n^{(s+1)}| \leqslant |e_{n-1}^{(s)} e_n^{(s)}|.$$

The quantity $|e_{n-1}^{(s)} e_n^{(s)}|$ is thus monotonic decreasing to either a non-zero limit L or to zero. In the latter case, we must eventually reach a state where either $|e_{n-1}^{(s)}| < \epsilon$ or $|e_n^{(s)}| < \epsilon$ for any $\epsilon > 0$. But since $|e_n^{(s+1)}| \leqslant |e_{n-1}^{(s)}|$ it follows that we must eventually reach the state when $|e_n^{(s)}| \leqslant \epsilon$. (In practice of course we may accept an $e_{n-1}^{(s)}$ which is negligible to working accuracy and eliminate the need for a further step of the QR algorithm.)

Suppose instead that

$$|e_{n-1}^{(s)} e_n^{(s)}| \to L > 0.$$

It follows that

$$\frac{|e_{n-1}^{(s+1)} e_n^{(s+1)}|}{|e_{n-1}^{(s)} e_n^{(s)}|} \to 1$$

and thus from relation (4), the quantities

$$|c_{n-2}^{(s)}|, \; |s_n^{(s)}|, \; |s_{n-1}^{(s)}|, \; \left| \frac{e_n^{(s)}}{\hat{d}_{n-1}^{(s)}} \right|$$

all tend to unity, implying that

$$|c_n^{(s)}| \to 0, \; |c_{n-1}^{(s)}| \to 0.$$

Thus

$$(s_n^{(s)})^2 = \frac{(e_n^{(s)})^2}{(e_n^{(s)})^2 + (x_{n-1}^{(s)})^2} \to 1$$

forcing $x_{n-1}^{(s)}$ to zero as $e_n^{(s)} \neq 0$ if $|e_n^{(s)}/\hat{d}_{n-1}^{(s)}|$ tends to unity. Now

$$x_{n-1}^{(s)} = \hat{d}_{n-1}^{(s)} c_{n-1}^{(s)} - e_{n-1}^{(s)} c_{n-2}^{(s)} s_{n-1}^{(s)}$$

and since $c_{n-1}^{(s)} \to 0$ and $x_{n-1}^{(s)} \to 0$ and $\hat{d}_{n-1}^{(s)}$ is bounded, by Lemma 10.1, it follows that

$$e_{n-1}^{(s)} c_{n-2}^{(s)} s_{n-1}^{(s)} \to 0.$$

But we have already shown that

$$|c_{n-2}^{(s)}| \to 1, \; |s_{n-1}^{(s)}| \to 1.$$

Hence $e_{n-1}^{(s)}$ tends to zero. This is a contradiction as $|e_{n-1}^{(s)} e_n^{(s)}| \to L$ and $e_n^{(s)}$ is bounded above. It follows that $|e_{n-1}^{(s)} e_n^{(s)}| \to 0$ and the convergence of the QR process is guaranteed.

Let us now turn to the question of rate of convergence. We prove first of all the following:

Lemma 10.2 If B is a real symmetric matrix and

$$QB = R$$

where Q is orthogonal and R is upper triangular then

$$|r_{nn}| \geqslant \min|\lambda_i(B)|.$$

Proof By definition, if $\|\mathbf{x}\|_2 = 1$,

$$\begin{aligned} \min|\lambda_i(B)| &= \min\|B\mathbf{x}\|_2 \leqslant \|BQ^T\mathbf{e}_n\|_2 \\ &= \|R\mathbf{e}_n\|_2 = |r_{nn}|. \end{aligned}$$

Let $\delta = \min|\lambda_i - \lambda_j| > 0$ where λ_i are the eigenvalues of A_1. Suppose we have reached the stage at which $e_n^{(s)} = \epsilon$ so that the shifted A_s is of the form

$$\begin{bmatrix} B & \epsilon\mathbf{e}_{n-1} \\ \epsilon e_{n-1}^I & \epsilon^2/\hat{d}_{n-1}^{(s)} \end{bmatrix}.$$

Let $\mu_1, \ldots, \mu_{n-1}$ be the eigenvalues of B and $\theta_1, \ldots, \theta_n$ be the eigenvalues of $A_s - k_s I$. Then for some ordering of the θ_i we may write

$$|\theta_i - \mu_i| \leqslant \epsilon, \; \left|\theta_n - \frac{\epsilon^2}{\hat{d}_{n-1}^{(s)}}\right| \leqslant \epsilon.$$

Now

$$\mu_i = (\mu_i - \theta_i) + (\theta_i - \theta_n) + \left(\theta_n - \frac{\epsilon^2}{\hat{d}_{n-1}^{(s)}}\right) + \frac{\epsilon^2}{\hat{d}_{n-1}^{(s)}}$$

giving the bound

$$\begin{aligned} |\mu_i| &\geqslant |\theta_i - \theta_n| - |\mu_i - \theta_i| - \left|\theta_n - \frac{\epsilon^2}{\hat{d}_{n-1}^{(s)}}\right| - \left|\frac{\epsilon^2}{\hat{d}_{n-1}^{(s)}}\right| \\ &\geqslant \delta - \epsilon - \epsilon - \frac{\epsilon^2}{|\hat{d}_{n-1}^{(s)}|} \\ &\geqslant \delta - 3\epsilon \end{aligned}$$

since $|\hat{d}_{n-1}^{(s)}| \geqslant \epsilon$ and the θ_i have the same separation as the λ_i.

Applying the first step of the QR transformation we eventually reach the stage when the configuration in the bottom two rows is

$$\begin{bmatrix} x_{n-1}^{(s)} & \epsilon c_{n-1}^{(s)} \\ \epsilon & \epsilon^2/\hat{d}_{n-1}^{(s)} \end{bmatrix}$$

where, by Lemma 10.2

$$|x_{n-1}^{(s)}| \geqslant \delta - 3\epsilon.$$

If we carry out the remainder of the transformation we find

$$e_n^{(s+1)} = \left(\frac{\epsilon^2 c_n^{(s)}}{\hat{d}_{n-1}^{(s)}} - \epsilon c_{n-1}^{(s)} s_n^{(s)}\right) s_n^{(s)}$$

where

$$|s_n^{(s)}| = \frac{\epsilon}{[(x_{n-1}^{(s)})^2 + \epsilon^2]^{1/2}} \leqslant \frac{\epsilon}{\delta - 3\epsilon}$$

giving finally

$$\begin{aligned} |e_n^{(s+1)}| &\leqslant \frac{\epsilon^3}{|\hat{d}_{n-1}^{(s)}|(\delta - 3\epsilon)} + \frac{\epsilon^3}{(\delta - 3\epsilon)^2} \\ &\leqslant \frac{\epsilon^2}{\delta - 3\epsilon} + \frac{\epsilon^3}{(\delta - 3\epsilon)^2} \end{aligned}$$

since

$$\epsilon \leqslant |\hat{d}_{n-1}^{(s)}|.$$

This result shows that convergence is certainly quadratic and hopefully cubic in general. The rate of convergence obviously depends on the behaviour of $\hat{d}_{n-1}^{(s)}$. We can see that the convergence is ultimately cubic by noting that for sufficiently small ϵ

$$|s_n^{(s)}| \to 0$$

$$|c_n^{(s)}| \to 1$$

and thus $|x_n^{(s)}|$ tends to $|\hat{d}_n^{(s)}|$ by equation (2). However by Lemma 10.2 we have that

$$|x_n^{(s)}| \geqslant \min|\lambda_i(A_1)|$$

so that

$$|\hat{d}_n^{(s)}| \geqslant |\min|\lambda_i(A_1)| - 0(\epsilon)|.$$

But

$$|\hat{d}_{n-1}^{(s)}| \geqslant |\hat{d}_n^{(s)}|,$$

and thus

$$|\hat{d}_{n-1}^{(s)}| \geqslant |\min|\lambda_i(A_1)| - 0(\epsilon)|.$$

In practice the regime of cubic convergence is reached rapidly.

During the course of the computation it is possible that some of the $|e_k^{(s)}|$ may become negligible to working accuracy. If this happens the matrix may be split into matrices of smaller order to improve the convergence of the algorithm.

If the eigenvectors of A_1 are also required the transformations will require to be stored. Notice that if A_1 is obtained from a general symmetric matrix by preliminary reduction to tridiagonal form then a double back transformation is required to obtain the eigenvectors.

The QR algorithm with the shift strategy outlined above is the most powerful algorithm for determining the eigenvalues of a symmetric tridiagonal matrix. In practice, seldom more than two iterations per eigenvalue are required.

Exercises

5. Apply the QR process given above to the 3×3 matrix

$$\begin{bmatrix} 0 & 1 & 0 \\ 1 & \epsilon & \epsilon \\ 0 & \epsilon & \epsilon \end{bmatrix}$$

where ϵ is a small number. Show that after the first step (whose shift is zero) e_3 is $0(\epsilon^2)$. Show however that after another QR step e_3 is $0(\epsilon^5)$.

6. Consider the shifted QR process with the shift strategy $k_s = d_n^{(s)}$. Show that

$$|e_n^{(s+1)}| \leqslant |e_n^{(s)}|$$

and that if $e_n^{(s)}$ does not tend to zero then $e_{n-1}^{(s)}$ does tend to zero.
(Hint: for the case when $|e_n^{(s)}| \rightarrow L > 0$ show that for sufficiently large s, $|s_n^{(s)}|$ and $|c_{n-1}^{(s)}|$ both tend to unity and thus $|s_{n-1}^{(s)}| \rightarrow 0$.)

7. Find the eigenvalues of the matrix

$$\begin{bmatrix} 2 & 1 & 0 \\ 1 & 2 & 1 \\ 0 & 1 & 2 \end{bmatrix}$$

by using (*a*) the QR process with shifts and (*b*) the QR process without shifts.

11

Extensions of Jacobi's method

11.1 Introduction

The reduction of a Hermitian matrix to a real diagonal matrix by a sequence of unitary similarity transformations is the basis of the Jacobi method described in Chapter 7. An important aspect of the method is that the eigenvalues and eigenvectors are obtained simultaneously, and an orthogonal set of eigenvectors is automatically produced even in the case of multiple eigenvalues. Further, the many variants of the method are extremely accurate when implemented on a digital computer. It is natural, therefore, to ask whether the procedure can be extended in some satisfactory way to non-Hermitian matrices.

Some attempts to answer this question are contained in this chapter. We begin by considering the class of matrices to which Jacobi-like processes would be expected to apply most naturally, that is the class of matrices which are unitarily similar to real *or complex* diagonal matrices. Such matrices are called *normal matrices*, and they have the advantage that the eigenvalue problem is always well-conditioned. In fact equation (27), Chapter 2, can be shown to hold for all normal matrices.

11.2 Normal matrices

If A is a normal matrix, then by definition there exists a unitary matrix R such that

$$R^*AR = \text{diag}(\lambda_i) = D, \text{ say.}$$

Thus

$$A = RDR^*,$$

and it follows that

$$A^*A = AA^*. \tag{1}$$

Further, we can write

$$A = B+C$$

where

$$B = \tfrac{1}{2}(A+A^*)$$

and

$$C = \tfrac{1}{2}(A-A^*).$$

The matrix B is Hermitian, and if the usual Jacobi procedure is applied directly, we will obtain

$$UBU^* = \text{diag}\,(\mu_i) = D_B, \text{ say,}$$

where U is unitary. We can write

$$B\mathbf{u}_i = \mu_i\mathbf{u}_i, \quad i = 1, 2, \ldots, n,$$

where $\mathbf{u}_i$ is the ith column of U^*, and so

$$BC\mathbf{u}_i = \mu_i C\mathbf{u}_i, \quad i = 1, 2, \ldots, n,$$

where we have used the fact that B and C commute. Thus, if $C\mathbf{u}_i$ is non-zero, it is an eigenvector of B corresponding to the eigenvalue μ_i, and this can be expressed as a linear combination of those eigenvectors $\mathbf{u}_j$ corresponding to μ_i. In particular, if the eigenvalues μ_i are distinct, then we have

$$C\mathbf{u}_i = \nu_i\mathbf{u}_i, \quad i = 1, 2, \ldots, n,$$

where the ν_i are constants, and in this case

$$C' = UCU^* = \text{diag}\,(\nu_i),$$

and A has been diagonalized. Otherwise, the only non-zero off-diagonal elements of C' can be in positions (i, j), where $\lambda_i = \lambda_j$. In this case, C' can readily be diagonalized by a sequence of plane rotations. The rotation matrices can be shown to commute with D_B, and thus leave it invariant, and so the diagonalization of A is again complete.

The extension of the usual Jacobi procedure to normal matrices has been considered in detail by Goldstine and Horwitz (1959). They develop a procedure in which $A = B+C$ is directly reduced to diagonal form by a sequence of unitary transformations. The method is optimal in the sense that in each step the sum of the absolute squares of the off-diagonal elements is decreased by as much as possible. The procedure outlined above is not necessarily optimal in this sense when B has multiple eigenvalues.

We consider the effect of operating on $A = B+C$ with a unitary matrix U defined as follows:

$$\begin{aligned} u_{ij} &= \delta_{ij}, \quad i, j \neq p, q \\ u_{pp} &= u_{qq} = \cos\theta \\ u_{pq} &= e^{i\alpha}\sin\theta \\ u_{qp} &= -e^{-i\alpha}\sin\theta \end{aligned}$$

for given p, q, α, θ.

Let $A' = B'+C'$, where $A' = UAU^*$, and B' and C' are similarly defined. Then for arbitrary $p \neq q$, the elements of A', B' and C' can readily be expressed in terms of those of A, B and C respectively.

Defining

$$\tau^2(H) = \sum_{i \neq j} |h_{ij}|^2 \tag{2}$$

for an arbitrary matrix H, we have

$$\tau^2(A') = \tau^2(A) + \Delta\tau^2(A),$$

where

$$\tfrac{1}{2}\Delta\tau^2(A) = |b'_{pq}|^2 + |c'_{pq}|^2 - |b_{pq}|^2 - |c_{pq}|^2.$$

Goldstine and Horwitz consider in some detail the problem of minimizing $\Delta\tau^2(A)$ with respect to the variables α and θ, assuming that p and q are fixed. They also show how to choose the pairs (p, q) in such a way that the process is convergent. The reader is referred to their paper for further details.

Exercises

1. Prove that $A^*A = AA^*$ if and only if A is normal. (This is sometimes used to define normality.)

2. Show that if $A = B + C$, as defined above,

$$\tau^2(A) = \tau^2(B) + \tau^2(C).$$

11.3 General matrices

In Chapter 2, we introduced the classical theorem of Schür, which states that an arbitrary complex matrix is unitarily similar to a triangular matrix, with the eigenvalues on the diagonal. Algorithms based on the reduction of a general matrix to triangular form by a sequence of plane rotations have been proposed by Greenstadt (1955) and Lotkin (1956). However both methods encounter difficulties: convergence has not been proved, nor have numerical experiments been particularly satisfactory.

The most successful approach to the Jacobi-like treatment of general matrices would appear to be that due to Eberlein (1962). The method is motivated by a theorem due to L. Mirsky which states that for all non-singular matrices H, values of $\|H^{-1}AH\|_E^2$ approach arbitrarily closely (from above) to $\sum_{j=1}^n |\lambda_j|^2$, where λ_j are the eigenvalues of A, with

$$\|H^{-1}AH\|_E^2 = \sum_{j=1}^{n} |\lambda_j|^2$$

for some H if and only if A is normal. Thus it follows that there exists a non-singular matrix H such that $A_L = H^{-1}AH$ is *arbitrarily close to being normal.* Using exercise 1 of Section 11.2, we see that this implies that the absolute value of every element of $(A_L A_L^* - A_L^* A_L)$ is arbitrarily small.

The matrix H is generated by forming the product of a sequence of matrices $H_j(p, q)$. Each H_j has the form

$$\begin{aligned} h_{pp} &= e^{-i\beta}\cos z \\ h_{pq} &= -e^{i\alpha}\sin z \\ h_{qp} &= e^{-i\alpha}\sin z \\ h_{qq} &= e^{i\beta}\cos z \\ h_{jk} &= \delta_{jk}, \quad j, k \neq p, q, \end{aligned} \tag{3}$$

where α, β are real, and $z = x+iy$, $i = \sqrt{-1}$. At each step of the iteration, the parameters which determine H_j may be chosen to minimize $\|H_j^{-1}AH_j\|_E^2$. The analysis is extremely tedious, and we merely give explicit approximations to the optimal parameters. Since these parameters are defined by two simultaneous quartic equations, their exact values are not in general available. For further details, the reader is referred to Eberlein's paper.

Let $A_0 = A$, normalized so that $\|A\|_E^2 \leqslant 1$. Let

$$A_{j+1} = H_j^{-1}A_jH_j, \quad j \geqslant 1,$$

where $H_j(p_j, q_j) = R_jS_j$ (where R_j and S_j are of the form (3)), and the pair (p_j, q_j) is chosen so that the absolute value of

$$4|w_{p_j, q_j}^{(j)}|^2 + (w_{p_j, p_j}^{(j)} - w_{q_j, q_j}^{(j)})^2$$

is at least equal to the average of such quantities for all possible p, q, where

$$W^{(j)} = A_jA_j^* - A_j^*A_j = U^{(j)} + iV^{(j)}.$$

Using subscripts R and S to distinguish the parameters corresponding to R and S respectively, and omitting the subscripts and superscripts j, the parameters are defined as follows:

$$R: \qquad \tan 2x_R = \frac{-w_{pp}+w_{qq}}{2u_{pq}}, \qquad \alpha_R = \beta_R = y_R = 0$$

$$S: \quad \tan(\alpha_S - \beta_S) = -\frac{1}{2}\frac{(2u_{pq})\cos 2x_R - (w_{pp}-w_{qq})\sin 2x_R}{v_{pq}}$$

$$\tanh y_S = \frac{\frac{1}{2}\sin(\alpha_S-\beta_S)\{(2u_{pq})\cos 2x_R - (w_{pp}-w_{qq})\sin 2x_R\} - \cos(\alpha_S-\beta_S)\,v_{pq}}{g_{pq}+2(|t_{pq}|^2+|d_{pq}|^2)}$$

where

$$g_{pq} = \sum_{j \neq p, q} \{|a_{pj}^2| + |a_{jp}^2| + |a_{qj}^2| + |a_{jq}^2|\},$$

$$t_{pq} = ((a_{pq} + a_{qp}) \cos 2x_R - (a_{pp} - a_{qq}) \sin 2x_R) \cos (\alpha_S - \beta_S)$$
$$- i(a_{pq} - a_{qp}) \sin (\alpha_S - \beta_S)$$

$$d_{pq} = (a_{pp} - a_{qq}) \cos 2x_R + (a_{pq} + a_{qp}) \sin 2x_R,$$

β_S and x_S are arbitrary.

For these choices, Eberlein shows that

$$\lim_{j \to \infty} \|W^{(j)}\|_E^2 = 0,$$

i.e. for sufficiently large j, A_j is arbitrarily close to being normal.

A simpler modification of this procedure applicable to *real matrices* is given by Eberlein, which reduces to the usual Jacobi procedure (or to the optimal procedure described in Section 11.2) when A is normal. Although there is no proof of convergence, successful results have been obtained. This particular strategy is such that the matrix R is chosen to minimize

$$\sum_{j \neq k} (a_{jk} + a_{kj})^2$$

and the matrix S to minimize $\|A\|_E^2$. R and S are both of the form (3), and the parameters are as follows:

$$R: \quad \tan 2x_R = \frac{a_{pq} + a_{qp}}{a_{pp} - a_{qq}}, \qquad \alpha_R = \beta_R = y_R = 0$$

$$S: \quad \tanh y_S = \frac{2[(w_{pp} - w_{qq}) \sin 2x_R - 2w_{pq} \cos 2x_R]}{g_{pq} + 2\{(a_{pq} - a_{qp})^2 + d_{pq}^2\}},$$

$$\beta_S = x_S = 0, \quad \alpha_S = -\pi/2,$$

where g_{pq} and d_{pq} are as before.

The limiting matrix in this case is partly diagonal and partly block diagonal with 2×2 blocks of the form

$$\begin{bmatrix} a_{mm} & a_{m,m+1} \\ -a_{m,m+1} & a_{mm} \end{bmatrix}.$$

Thus the eigenvalues can be read off directly.

As in the classical Jacobi method, the method has the advantage that only one program is necessary to give the complete eigensystem. Although no error analysis exists, experimental evidence indicates that very good results are in general obtained, and it is possible that further development could lead to an algorithm competitive with the QR algorithm (see Chapter 13), particularly for small matrices.

12

Extensions of Givens' and Householder's methods

12.1 Introduction

The procedures outlined in Chapter 11, based on extensions of the Jacobi method for Hermitian matrices, are somewhat cumbersome and, in the main, unproven techniques, though it is possible that a powerful algorithm may develop from that approach.

In this chapter, we turn to consider generalizations of the methods of Givens and Householder to reduce the original matrix to simpler form, and of the Sturm sequence bisection routine for determining the eigenvalues. As we shall see the first of these goals is easily achieved, the other requires considerably more thought. The extension of the QR process (Chapter 10) which proved so effective for the symmetric tridiagonal problem, is considered in the next chapter.

12.2 Reduction to upper Hessenberg form

A Hermitian matrix may be unitarily reduced to symmetric tridiagonal form by means of the Givens or Householder process. If A is a general complex matrix we can no longer hope to transform it into as simple a form. In fact, our aim will be to reduce a general matrix A to an upper Hessenberg matrix. In this situation, it is not obvious that a unitary reduction by means of either the Givens or the Householder process is preferable to a straightforward similarity reduction based on elementary operation matrices. The latter technique is frequently employed in practice though there are possible dangers. These we will discuss later.

A reduction based on elementary unitary matrices is defined in an entirely analogous way to that which applies to Hermitian matrices. However we can no longer invoke symmetry to reduce the number of operations to be carried out. All that we can assume is that the set of elements we have chosen to annihilate may be immediately set to zero and thus need not be computed. Moreover we may always carry through a reduction to an upper Hessenberg matrix with real subdiagonal elements by means of the P_λ matrices outlined in Section 3.4, or by means of elementary unitary transformations followed by further diagonal unitary transformations. The development of these processes we leave as an exercise.

As mentioned above it is possible to generate a similarity reduction of a general complex matrix to upper Hessenberg form by means of the elementary operation matrices discussed in Chapter 3. The procedure will be similar to that employed in the Gaussian elimination algorithm with the additional requirement of the completion of the similarity transformation. The use of interchanges will again play an important part in the construction of the process.

To illustrate the method consider the reduction of a 5×5 matrix to upper Hessenberg form. We assume that we have reached the stage when the appropriate elements in the first column have already been reduced to zero. Thus let A be of the form

$$\begin{bmatrix} x & x & x & x & x \\ x & x & x & x & x \\ 0 & x & x & x & x \\ 0 & \boxed{x} & x & x & x \\ 0 & \boxed{x} & x & x & x \end{bmatrix}.$$

In this stage of the algorithm, we intend to use simple row operations to reduce the boxed elements to zero, in such a way that completion of the similarity transformation will not affect the zeros in the first column *or* the zeros which have just been introduced in the second column. Our row operations will involve rows 3 to 5 in order to reduce elements 4 and 5 of column 2 to zero. (More generally we will use row operations on rows $r+1$ to n in order to reduce elements $r+2$ to n of column r to zero.)

The first step is to determine that element amongst elements 3 to 5 of column 2 which has largest modulus (the pivot search). Suppose this is element 5 in the column. We interchange rows 3 and 5 and by means of the usual technique subtract multiples of row 3 from rows 4 and 5 in order to reduce the (4, 2), (5, 2) elements to zero. The matrix now has the form

$$\begin{bmatrix} x & x & x & x & x \\ x & x & x & x & x \\ 0 & a & a & a & a \\ 0 & 0 & a & a & a \\ 0 & 0 & a & a & a \end{bmatrix}$$

where 'a' denotes an element which has been modified. These row operations are achieved by the elementary matrix $S_{ij}(k)$ discussed in Section 3.2. The next step is to interchange columns 3 and 5, and in order to complete the similarity transformation, we will require $S_{ij}^{-1}(k)$. In exercise 1 of Section 3.2 it was shown that

$$S_{ij}^{-1}(k) = S_{ij}(-k).$$

Thus corresponding to a subtraction of a multiple of row 3 from row 4, say, achieved by premultiplication by $S_{43}(k)$, postmultiplication by $S_{43}(-k)$ will require the addition of the same multiple of column 4 to column 3. Of course in practice, the annihilation of a column is carried out completely and the completion of the similarity transformation only then carried out. This process does not affect any element reduced to zero by the premultiplication. We therefore obtain a matrix of the form

$$\begin{bmatrix} x & x & a & a & a \\ x & x & a & a & a \\ 0 & a & a & a & a \\ 0 & 0 & a & a & a \\ 0 & 0 & a & a & a \end{bmatrix}.$$

One further similarity transformation will reduce the (5, 3) element to zero, giving an upper Hessenberg matrix.

If the arithmetic is organized efficiently, then this similarity reduction may be accomplished in approximately $\frac{5}{6}n^3$ multiplications. This is half the number required by the Householder process which is itself half that required by the Givens process. However, the unitary reduction is much more stable as an example will show.

Consider the 6×6 matrix

$$\begin{bmatrix} 0 & 1 & 1 & 1 & 1 & 1 \\ 1 & 0 & 0 & 0 & 0 & -1 \\ -1 & 1 & 0 & 0 & 0 & -1 \\ -1 & 0 & 1 & 0 & 0 & -1 \\ -1 & 0 & 0 & 1 & 0 & -1 \\ 0 & 0 & 0 & -\frac{1}{2} & \frac{1}{2} & 0 \end{bmatrix}.$$

This may be reduced to the Hessenberg form

$$\begin{bmatrix} 0 & -2 & -1 & 0 & 0 & 1 \\ 1 & 0 & 0 & 0 & 1 & -1 \\ & 1 & 0 & 0 & 2 & -2 \\ & & 1 & 0 & 4 & -4 \\ & & & 1 & 8 & -8 \\ & & & & 8\frac{1}{2} & -8 \end{bmatrix}$$

by the above process. If, on computation, one restricts the word length to four bits then the above matrix is not obtained, though the only modification is to replace the (6, 5) element by 8. The eigenvalues of this perturbed matrix are considerably modified, and the reduction is not stable. Of course, there can be no question about the stability of a unitary Householder reduction. (The above example is due to Businger.)

Thus a general recommendation would be to use a Householder reduction to upper Hessenberg form.

Exercises

1. Show that it is possible to produce a similarity reduction of a general matrix to upper Hessenberg form giving subdiagonal elements with the value one.

2. Use the various techniques discussed in this section to reduce to upper Hessenberg form the matrix

$$\begin{bmatrix} 1 & 2 & 2 & 0 \\ 3 & 1 & 3 & 3 \\ 2 & 2 & 1 & 2 \\ 1 & 2 & 3 & 1 \end{bmatrix}.$$

12.3 Further reduction to tridiagonal form

It is not possible to reduce the upper Hessenberg matrix to the simpler tridiagonal form by means of elementary unitary matrices. However, it is possible, at least in theory, to use elementary operation matrices to provide a similarity transformation to tridiagonal form. Unfortunately, the use of interchanges is not allowed in this further reduction. In this sense, it is equivalent to applying the above elimination process to reduce a lower Hessenberg matrix to upper Hessenberg (actually tridiagonal) form. Thus consider the case of a 5×5 matrix at the stage

$$\begin{bmatrix} x & x & 0 & 0 & 0 \\ x & x & x & 0 & 0 \\ 0 & x & x & x & 0 \\ 0 & \boxed{x} & x & x & x \\ 0 & \boxed{x} & x & x & x \end{bmatrix}.$$

The two boxed elements are made to vanish by subtracting multiples of the third row from their respective rows. This of course is not possible if $a_{32} = 0$, which corresponds to a case where we would employ interchanges. Since we are not allowed to pivot the process breaks down. Notice that this part of the elimination process does not affect any existing zero elements.

The matrix will now be of the form

$$\begin{bmatrix} x & x & 0 & 0 & 0 \\ x & x & x & 0 & 0 \\ 0 & x & x & x & 0 \\ 0 & 0 & x & x & x \\ 0 & 0 & x & x & x \end{bmatrix}.$$

To complete the similarity reduction we *add* the appropriate multiples of columns 4 and 5 to column 3, the multipliers being those defined in the first stage. Obviously, this again does not introduce any new non-zero elements.

The above process can produce catastrophic results and *is not recommended.* In fact, while there are other techniques for the tridiagonalization of a general matrix, their possible instabilities cannot be too strongly stressed. If the matrix A is not Hermitian, then one should be content with a reduction to upper Hessenberg form. Therefore, the remainder of this chapter will be devoted to discussing the computation of the eigensystem of an upper Hessenberg matrix.

12.4 Evaluation of the characteristic polynomial

The solution of the eigenproblem for a tridiagonal matrix benefited by the existence of the Sturm sequence property. Unfortunately no such simple algorithm exists for a complex Hessenberg matrix which will give us detailed information on the distribution of the eigenvalues. The determination of the eigenvalues of a matrix is akin to the determination of the roots of a polynomial equation. It is not surprising therefore that the methods we will shortly discuss were originally constructed for this latter purpose. The main difference in approach is that we do not have an explicit expression for the characteristic polynomial. Our approach will involve the evaluation of the polynomial and its derivatives (but notice not the evaluation of its coefficients) by an indirect method. *The great majority of procedures which attempt to construct the characteristic polynomial explicitly are unstable and should not be used.*

The following indirect method, usually referred to as Hyman's method, is used to compute the value of the characteristic polynomial of an upper Hessenberg matrix. Let H denote an upper Hessenberg matrix of order n. We consider the set of equations

$$(H-\lambda I)\,\mathbf{x} = k(\lambda)\,\mathbf{e}_1 \tag{1}$$

where $\mathbf{x}$ is an n-vector and where

$$\mathbf{e}_1^T = (1, 0, \ldots, 0)$$

with $k(\lambda)$ a constant. By arbitrarily choosing x_n equal to one we may evaluate $x_{n-1}, \ldots, x_2, x_1$ from the last $(n-1)$ equations in (1) for a particular choice of λ. The first equation in (1) may now be used to determine $k(\lambda)$. By considering this process for a general x_n we observe that $k(\lambda)$ is an nth degree polynomial in λ with coefficients depending only on x_n. For a value of λ such that $k(\lambda) = 0$ it follows that $(H-\lambda I)\,\mathbf{x} = \mathbf{0}$. There are at most n such values of λ since $k(\lambda)$ is an nth degree polynomial (for a fixed x_n). Therefore $k(\lambda)$ must be proportional through a constant factor (dependent on x_n) to the characteristic polynomial $f(\lambda)$ which is equal to the determinant of $(H-\lambda I)$.

Formally, let Q denote the matrix $H-\lambda I$ and let the columns of Q be denoted by $\mathbf{q}_1, \ldots, \mathbf{q}_n$. The system (1) requires the determination of $x_n(= 1), x_{n-1}, \ldots, x_1$ such that we may determine $k(\lambda)$ with

$$x_1\mathbf{q}_1 + \ldots + x_{n-1}\mathbf{q}_{n-1} + \mathbf{q}_n = k(\lambda)\,\mathbf{e}_1. \tag{2}$$

We may regard (2) as defining $\mathbf{q}_n$ in terms of $\mathbf{q}_1, \ldots, \mathbf{q}_{n-1}$ and $\mathbf{e}_1$ so that

$$\mathbf{q}_n = k(\lambda)\mathbf{e}_1 - \sum_{i=1}^{n-1} x_i\mathbf{q}_i.$$

It is well-known that if we modify one column of a matrix by adding linear combinations of the other columns of the matrix to it, the value of the determinant of the matrix is unchanged. Thus, the determinant of $Q = H-\lambda I$ is equal to the determinant of the matrix Q with its final column replaced by $k(\lambda)\mathbf{e}_1$, that is the matrix

$$[\mathbf{q}_1, \ldots, \mathbf{q}_{n-1}, k(\lambda)\mathbf{e}_1]. \tag{3}$$

However, since Q is an upper Hessenberg matrix, the matrix (3) has the form (in the 5×5 case as an example)

$$\begin{bmatrix} x & x & x & x & x \\ x & x & x & x & 0 \\ 0 & x & x & x & 0 \\ 0 & 0 & x & x & 0 \\ 0 & 0 & 0 & x & 0 \end{bmatrix}.$$

Expanding the determinant of (3) by the final column gives the result

$$\det(H-\lambda I) = f(\lambda) = (-1)^{n+1}k(\lambda)h_{21}h_{32}, \ldots, h_{n,\,n-1}.$$

For any value of λ the value of the determinant of $H-\lambda I$ is proportional through a *constant* factor to $k(\lambda)$. It follows that the eigenvalues of H which are of course the zeros of $f(\lambda)$ are also the zeros of $k(\lambda)$. (We assume that no $h_{j,j-1}$ is zero for otherwise we could work with matrices of lower order.) Moreover, it is possible to produce a reduction to upper Hessenberg form which gives subdiagonal elements with the value one (see exercise 1 of Section 12.2). In this case, the values of $|f(\lambda)|$ and $|k(\lambda)|$ are equal.

It is therefore possible, even in general, to deal with values of $k(\lambda)$ as opposed to those of $f(\lambda)$ in order to find the eigenvalues. Writing out the system (1) gives

$$\begin{aligned} (h_{11}-\lambda)\,x_1+h_{12}x_2+\ldots\ldots\ldots\ldots+h_{1n}x_n &= k(\lambda) \\ h_{21}x_1+(h_{22}-\lambda)\,x_2+\ldots\ldots\ldots+h_{2n}x_n &= 0 \\ h_{32}x_2+(h_{33}-\lambda)\,x_3+\ldots+h_{3n}x_n &= 0 \\ &\;\,\vdots \\ h_{n,\,n-1}x_{n-1}+(h_{nn}-\lambda)\,x_n &= 0 \end{aligned} \tag{4}$$

which we solve backwards starting with $x_n = 1$. This is an efficient and stable way of computing $k(\lambda)$.

Some of the techniques to be discussed may also require values of the derivatives of $f(\lambda)$. These may be obtained in an entirely analogous manner. Remembering that the vector $\mathbf{x}$ is a function of λ we may differentiate (1) with respect to λ to give

$$(H-\lambda I)\,\mathbf{x}'-\mathbf{x} = k'(\lambda)\,\mathbf{e}_1 \tag{5}$$

where the prime denotes differentiation with respect to λ. Since the elements of $\mathbf{x}$ for a particular λ may be computed by the system (4) and since $x_n = 1$ implies $x_n' = 0$ the system (5) may likewise be used to compute the values of

$x'_{n-1}, \ldots, x'_1, k'(\lambda)$. The value of $k'(\lambda)$ is again proportional (by the same constant of proportionality) to $f'(\lambda)$.

Higher derivatives of $k(\lambda)$ may be obtained in a similar manner by repeated differentiation. The computation may be efficiently organized by computing both function values and derivatives in the one backward sweep.

Notice that nowhere in this calculation do we obtain an *explicit* representation for $f(\lambda)$, or any of its derivatives. In theory, by evaluating $f(\lambda)$ for $n+1$ values of λ, we could solve the set of equations

$$a_n\lambda_i^n + a_{n-1}\lambda_i^{n-1} + \ldots + a_1\lambda_i + a_0 = f(\lambda_i), \quad i = 1, \ldots, (n+1)$$

to determine the coefficients a_i, but this is a dangerous process which, again, we do not recommend. In any case, we do not require an explicit form for $f(\lambda)$ or any of its derivatives in order to compute its zeros. It is sufficient to be able to find numerical values of $f(\lambda)$ and its derivatives *implicitly*.

Exercise

3. Use the above processes to evaluate $|f(\lambda)|$ and $|f'(\lambda)|$ for suitable values of λ for the matrix

$$\begin{bmatrix} 1 & 2 & 2 \\ 1 & 4 & 1 \\ 0 & 1 & 2 \end{bmatrix}$$

and check your answers by comparison with the true values.

12.5 Computation of the eigenvalues

The computation of the eigenvalues of an upper Hessenberg matrix, whose determinant is determined implicitly by Hyman's method, is exactly equivalent to finding the zeros of a polynomial equation. Since there is this equivalence any method which uses only values of $f(\lambda)$ and its derivatives (but not *explicit* coefficients from them) may be used to determine the eigenvalues of $H-\lambda I$. An excellent treatise on this subject is the book by Traub (1964). We will content ourselves with a brief description of a few of the possible approaches. The interested reader is referred to Wilkinson (1965) for fuller details on their application to the matrix eigenproblem.

We begin with a consideration of the bisection method used so successfully in Chapter 9. Here we apply the same type of strategy although now we use values of $f(\lambda)$ as opposed to those of $s(\lambda)$. (Note however that in the symmetric tridiagonal case we could have used the values of $p_n(\lambda)$ which correspond to $f(\lambda)$.) Our aim is initially to find an interval on the real λ line (we can only find real roots of $f(\lambda)$ by this approach) which is such that the signs of $f_1 = f(\lambda_1)$ and $f_2 = f(\lambda_2)$ are different. When this occurs it follows that there is at *least* one root of $f(\lambda)$ in the interval between λ_1 and λ_2. If the interval is not too wide we can hope that there will be not more than one root in it. By evaluating $f(\lambda)$ at $\lambda = \frac{1}{2}(\lambda_1+\lambda_2)$ we can determine a smaller interval for this root. The new interval is taken as that for which the sign of one of $f(\lambda_1), f(\lambda_2)$ differs from that of $f(\lambda)$. The width of the interval has been halved and it follows that successive use of this rule allows the process of bisection to locate an interval of small width in which the root lies. Notice however that we have no analogue of the Sturm sequence property to give us information on the distribution of eigenvalues. Therefore some difficulty may be experienced in finding a 'good' starting interval. Further, the method of bisection is restricted to determining real roots.

This drawback is also present if we use the linear interpolation formula

$$\lambda = \lambda_1 - f_1 \frac{(\lambda_1-\lambda_2)}{(f_1-f_2)}$$

with $f_1 = f(\lambda_1)$ and $f_2 = f(\lambda_2)$ and start with two real values λ_1, λ_2 of λ. (This is particularly common when finding the eigenvalues of a real upper Hessenberg matrix.)

Fortunately if we use a method based on quadratic interpolation then it is possible to start from real values of λ and f and move into the complex plane. This process of quadratic interpolation is usually known as Muller's method. Our first aim is to determine the coefficients a, b, c of the quadratic

$$\phi(\lambda) = a+b\lambda+c\lambda^2 \tag{6}$$

which passes through the points (λ_1, f_1), (λ_2, f_2), (λ_3, f_3). Then as a better approximation to the eigenvalue we will take a root of $\phi(\lambda) = 0$. The conditions for the quadratic to pass through these three points are given by

$$\begin{aligned} f_1 &= a+b\lambda_1+c\lambda_1^2 \\ f_2 &= a+b\lambda_2+c\lambda_2^2 \\ f_3 &= a+b\lambda_3+c\lambda_3^2. \end{aligned} \tag{7}$$

Introducing the divided difference notation

$$f[\lambda_3, \lambda_2, \lambda_1] = \frac{f[\lambda_3, \lambda_2]-f[\lambda_2, \lambda_1]}{\lambda_3-\lambda_1}$$

where

$$f[\lambda_2, \lambda_1] = \frac{f_2-f_1}{\lambda_2-\lambda_1},$$

the solution of (7) is given by

$$\begin{aligned} c &= f[\lambda_3, \lambda_2, \lambda_1] \\ b &= f[\lambda_3, \lambda_2]-(\lambda_3+\lambda_2)f[\lambda_3, \lambda_2, \lambda_1] \\ a &= f_3-\lambda_3 f[\lambda_3, \lambda_2]+\lambda_3\lambda_2 f[\lambda_3, \lambda_2, \lambda_1]. \end{aligned}$$

In fact, we shall not require the value of a. By subtracting the third equation of (7) from (6) we obtain

$$
\begin{aligned}
\phi(\lambda)-f_3 &= b(\lambda-\lambda_3)+c(\lambda^2-\lambda_3^2)\\
&= (\lambda-\lambda_3)\{f[\lambda_3, \lambda_2]-(\lambda_3+\lambda_2)f[\lambda_3, \lambda_2, \lambda_1]\}+(\lambda^2-\lambda_3^2)f[\lambda_3, \lambda_2, \lambda_1]\\
&= (\lambda-\lambda_3)\{f[\lambda_3, \lambda_2]+(\lambda_3-\lambda_2)f[\lambda_3, \lambda_2, \lambda_1]\}+(\lambda-\lambda_3)^2 f[\lambda_3, \lambda_2, \lambda_1]\\
&= (\lambda-\lambda_3)\,X+(\lambda-\lambda_3)^2\,Y
\end{aligned}
$$

in an obvious notation. It follows that the zeros of $\phi(\lambda)$ are given by

$$\lambda = \lambda_3-2f_3/\{X\pm[X^2-4f_3Y]^{1/2}\}. \tag{8}$$

The great merit of this formulation is that we need only evaluate $f[\lambda_3, \lambda_2]$ at each step. The process is iterative in that we now apply the same technique to the points λ_2, λ_3, λ_4. The sign in (8) is chosen to give λ_4 the smaller modulus of the two roots. Clearly the roots of (8) may be complex even if we start from real points.

In general, we do not use interpolation formulas of order greater than two. The extra effort required in solving a cubic (in cubic interpolation) may reduce any hoped-for increase in convergence rate. The above interpolation methods have only employed function values. Since it is just as straightforward to evaluate derivatives of $f(\lambda)$ when using Hyman's method it seems worthwhile considering the use of such methods to resolve the eigenvalue problem. However we limit ourselves to a brief discussion.

If we are prepared to evaluate both $f(\lambda)$ and $f'(\lambda)$ then we may use Newton's method in the form:

$$\lambda_{k+1} = \lambda_k-\frac{f(\lambda_k)}{f'(\lambda_k)}. \tag{9}$$

If the root is simple, then this method has quadratic convergence, that is

$$|\lambda_{k+1}-\lambda|\propto|\lambda_k-\lambda|^2.$$

However if the root to which the process is converging is multiple, then (9) has linear convergence. It is possible to modify the formula to take account of multiplicity of roots *if* we know the order of multiplicity of the root. In general, we will not have this information. Further, notice that if H is a real upper Hessenberg matrix, then both $f(\lambda)$ and $f'(\lambda)$ will be real quantities when computed by Hyman's method. Thus, like linear interpolation, the sequence of iterates produced by Newton's method is always real, if the initial guess is real, and therefore we cannot determine complex roots by this process.

A cubically convergent method which is frequently used in practice is that due to Laguerre. For an nth degree polynomial $f(\lambda)$ we compute at the current point λ_k the values of $f(\lambda_k), f'(\lambda_k), f''(\lambda_k)$ and then form

$$X^2 = (n-1)[(n-1)\{f'(\lambda_k)\}^2 - nf(\lambda_k)f''(\lambda_k)].$$

The next approximation is given by the formula

$$\lambda_{k+1} = \lambda_k - nf(\lambda_k)/(f'(\lambda_k) \pm X). \tag{10}$$

The sign is chosen so that the denominator has the larger modulus. Unfortunately, this process loses its cubic convergence rate if the sequence is tending to a multiple eigenvalue. Like Newton's method, we may modify (10) to maintain cubic convergence but this again requires that we be able to predict the multiplicity of the root, which is difficult to do in practice. Note that Laguerre's method, like Muller's method, allows the possibility of moving into the complex plane.

If the upper Hessenberg matrix H is real, then it follows that the characteristic polynomial of H has real coefficients and that the eigenvalues of H are either real or occur in complex conjugate pairs. Thus, if we obtain a complex eigenvalue of a real H we may immediately write down its conjugate as also being a root. There exist special methods which use a strategy based on the occurrence of complex conjugate pairs of roots. Their aim is to determine quadratic factors of the real polynomial $f(\lambda)$. We refer the interested reader to Wilkinson (1965) for further details on this topic and on the question of starting the iteration.

Having computed one or more of the eigenvalues by any of the above techniques, it is desirable to ensure that in further searches for eigenvalues, we do not redetermine those already found. Thus we require some technique of *suppressing* the known eigenvalues. The recommended technique is as follows. If approximations $\lambda^{(1)}, \ldots, \lambda^{(s)}$ to eigenvalues are known, then instead of iterating with $f(\lambda)$ use

$$g(\lambda) = f(\lambda)\Big/\prod_{i=1}^{s}(\lambda - \lambda^{(i)}). \tag{11}$$

Note that $g(\lambda)$ is again implicitly defined through the implicit determination of $f(\lambda)$. If we require derivatives of $g(\lambda)$ (for example to use in Newton's or Laguerre's method) these are easily obtained from (11). Thus, for example,

$$g'(\lambda) = g(\lambda)\left\{f'(\lambda)/f(\lambda) - \sum_{i=1}^{s}(\lambda - \lambda^{(i)})^{-1}\right\} \tag{12}$$

from which we obtain an expression for $g'(\lambda)/g(\lambda)$ required in Newton's method. Laguerre's method requires the second derivative of $g(\lambda)$. This is obtained by differentiation of (12) giving

$$\frac{g''}{g} = \frac{f''}{f} - \left(\frac{f'}{f}\right)^2 + \sum_{i=1}^{s}(\lambda - \lambda^{(i)})^{-2} + \left(\frac{g'}{g}\right)^2$$

where g'/g is computed (implicitly) by (11). Difficulties may arise when clusters of eigenvalues are present. Such problems are beyond the scope of this text.

Exercises

4. Under certain pathological circumstances, Laguerre's method can fail. Show that this is the case in determining the zeros of $z(z^2+b^2)$, $b>0$, taking $z_1 = b/\sqrt{3}$, by verifying that $z_2 = -z_1$, $z_3 = z_1$, etc.

5. Verify the quadratic (cubic) convergence rates of Newton's (Laguerre's) method by solving a low-order polynomial equation with simple roots. Repeat the process to find a multiple root of a low-order polynomial.

6. The determinant of a matrix may be evaluated as part of the usual Gaussian elimination process. Alternatively, we may evaluate it by reducing the matrix to upper Hessenberg form and applying Hyman's technique. Show how this is achieved.

7. Laguerre's method requires the determination of $k''(\lambda)$. Derive the equation similar to (5) to achieve this.

8. For the matrix given in exercise 3 (Section 12.4) determine the eigenvalues by each of the techniques discussed in this section. Use Hyman's method to evaluate the necessary quantities $k(\lambda)$, $k'(\lambda)$, $k''(\lambda)$.

9. For the example of the above matrix, show that for the eigenvalue $\lambda = 1$ of $(H-\lambda I)$ the vector $\mathbf{x}'$, defined in the Hyman process for $k'(\lambda)$ is an eigenvector of $(H-\lambda I)^2$. Why?

12.6 Evaluation of eigenvectors

The evaluation of an eigenvector belonging to an approximate eigenvalue of an upper Hessenberg matrix is achieved by inverse iteration described in Chapter 6. Notice that if we find that the matrix H has an eigenvalue of multiplicity m then there may be any number of eigenvectors p corresponding to it where

$$1 \leqslant p \leqslant m.$$

In general, we will be unable to determine the degree of degeneracy. This situation is totally different from that of a Hermitian matrix which has a full set of eigenvectors. However in practice it will be rare for the eigenvalues to be exactly equal, and trouble is more likely to arise from close eigenvalues with corresponding difficulty in obtaining distinct eigenvectors.

The eigenvectors of the original matrix may be determined from those of H by reversing the Householder reduction.

13

QR algorithm for Hessenberg matrices

13.1 Introduction

In Chapter 12, it was shown that an arbitrary complex matrix could be reduced to upper Hessenberg form by a similarity transformation. If the original matrix consisted entirely of real elements, then the Hessenberg matrix would also be real. Moreover, in the general case the transformation was unique if we insisted on the subdiagonal elements of the Hessenberg matrix being real and positive.

In this chapter, we consider algorithms for the determination of the complete eigensystem (that is eigenvalues, or eigenvalues *and* eigenvectors) of upper Hessenberg matrices, by means of extensions of the QR algorithm discussed in Chapter 10. Little will be said of a general LR process although it is of more value in this general situation. We will concentrate our attention on the QR process for a complex Hessenberg matrix and on the double QR process for a real Hessenberg matrix.

13.2 QR algorithm for a complex Hessenberg matrix

We consider first of all the eigenproblem for a complex Hessenberg matrix. We may assume that no subdiagonal element $a_{i+1,i}$ is zero as otherwise we could work with matrices of lower order.

The restoring QR algorithm with shifts is then given by

$$A_s - k_s I = Q_s R_s \tag{1}$$

$$A_s = k_s I + R_s Q_s$$

where Q_s is a unitary matrix and where R_s is an upper triangular matrix which is usually restricted to having real diagonal elements.

There is little difference between the form of discussion for the process (1) and that for the QR algorithm (1), Chapter 10, for a real symmetric tri-diagonal matrix. It may readily be verified that if A_s is of upper Hessenberg form then the transformation (1) is such as to ensure that A_{s+1} is of the same form. Moreover, our choice of shift k_s may be made according to the same rule discussed in Chapter 10. The complex shift k_s will now be the eigenvalue of the bottom right-hand 2×2 matrix of A_s which is closest in modulus to $a_{nn}^{(s)}$.

The convergence of the process (1) has been shown under general conditions but the proofs as yet do not give any indication of the rate of convergence. The proofs, which are given in detail in the book of Wilkinson (1965), are not repeated here. We content ourselves with an elementary discussion of the terminal properties of the general QR algorithm.

First of all we must define what we mean by 'convergence' of the QR algorithm. (This problem did not exist in the real symmetric case as the QR factorization was automatically unique.) We require to choose Q_s such that R_s is an upper triangular matrix with real diagonal elements. This ensures the uniqueness of the algorithm but has one drawback. Consider the QR process applied to an upper triangular matrix A_1. If some of the diagonal elements of A_1 are complex, it follows that $R_1 \neq A_1$. If $a_{jj} = |a_{jj}|\, e^{i\theta_j}$ and if we define the diagonal matrix D so that

$$D = \operatorname{diag}\{e^{i\theta_j}\} \quad (i = \sqrt{-1})$$

then the QR decomposition is given by

$$Q_1 = D, \quad R_1 = D^{-1}A_1,$$

and thus

$$A_2 = D^{-1}A_1D.$$

The diagonal elements of A_2 (which is of course still upper triangular) are the same as those of A_1 but the elements in the strict upper triangle have been multiplied by complex numbers of unit modulus. Obviously, either A_1 or A_2 would suffice to determine the eigenvalues but, in the normal sense, R_2 and R_1 do not belong to a convergent sequence. However, such diagonal transformations are of no importance to the determination of the eigenvalues of A_1. Thus we say that the *QR* process is *essentially convergent* if A_{s+1} tends to $D^{-1}A_sD$ for some diagonal unitary matrix D.

Wilkinson (1965) shows that the unshifted QR algorithm may under certain circumstances generate a sequence of matrices A_s which do not converge to an upper triangular matrix. However the *shifted* QR algorithm, except in rare pathological cases, essentially converges. In practice, the use of shifts is a prerequisite for a reasonable rate of convergence. Just as in the symmetric case, we use the deflation strategy to reduce the dimension of the matrices generated by the QR process. More involved deflating techniques are also available which make the QR algorithm a very powerful routine.

13.3 Double QR algorithm for a real Hessenberg matrix

If the Hessenberg matrix A_1 is real then one would ideally like to avoid the use of complex arithmetic and construct by means of the QR algorithm a sequence of matrices A_s which converge to a form from which the eigensystem may readily be determined. If it were known at the outset that all the eigenvalues of A_1 were real, then the use of suitably chosen real shifts would enable us to generate a sequence of A_s tending to real upper triangular form. In this case, all Q_s would be real and, moreover, essential convergence would be replaced by convergence.

However in this section we consider an ingenious algorithm due to Francis which finds the eigensystem of a general real Hessenberg matrix using real arithmetic only. Of course we can no longer expect the algorithm to generate a sequence of matrices A_s which tend to strict upper triangular form. The limiting form of the matrix A_s will be block upper triangular with 1×1 or 2×2 blocks on the diagonal. The 1×1 blocks will correspond to real eigenvalues and the 2×2 blocks to complex conjugate pairs of eigenvalues.

Consider a general step in the QR algorithm applied to the real Hessenberg matrix A_1. In determining the shift to be used for the following step, we find the eigenvalues of the 2×2 real matrix in the bottom right-hand corner of the matrix A_{2s}. Let these be $k_s^{(1)}$ and $k_s^{(2)}$. Either both $k_s^{(1)}$ and $k_s^{(2)}$ are real, or they form a complex conjugate pair. In the following discussion, which of these occurrences is the actual situation is irrelevant. Instead of taking a single step of the QR algorithm with the usual shift strategy let us consider the determination of $A_{2(s+1)}$ by means of two steps one of which uses $k_s^{(1)}$ and the other $k_s^{(2)}$. Thus

$$
\begin{aligned}
A_{2s} - k_s^{(1)} I &= Q_{2s} R_{2s} \\
A_{2s+1} - k_s^{(1)} I &= R_{2s} Q_{2s} \\
A_{2s+1} - k_s^{(2)} I &= Q_{2s+1} R_{2s+1} \\
A_{2(s+1)} - k_s^{(2)} I &= R_{2s+1} Q_{2s+1}.
\end{aligned}
\tag{2}
$$

From these relations we find that

$$(Q_{2s} Q_{2s+1})(R_{2s+1} R_{2s}) = (A_{2s} - k_s^{(1)} I)(A_{2s} - k_s^{(2)} I) \tag{3}$$

and

$$A_{2(s+1)} = (Q_{2s} Q_{2s+1})^* A_{2s} (Q_{2s} Q_{2s+1}). \tag{4}$$

Since $k_s^{(1)} + k_s^{(2)}$ and $k_s^{(1)} k_s^{(2)}$ are real, it follows that the right-hand side of (3) is real. Further since we have determined Q_{2s} and Q_{2s+1} so that R_{2s} and R_{2s+1} have real diagonal elements (thus ensuring uniqueness) it follows that the matrix $R_{2s+1} R_{2s}$ has real diagonal elements. But the triangularization of the real matrix

$$(A_{2s} - k_s^{(1)} I)(A_{2s} - k_s^{(2)} I) \tag{5}$$

is unique, and thus it follows that $R_{2s+1} R_{2s}$ is a real upper triangular matrix, and $Q_{2s} Q_{2s+1}$ a real orthogonal matrix. Thus from (4) we may achieve the transformation from A_{2s} to $A_{2(s+1)}$ by means of real arithmetic.

We could determine $Q_{2s}Q_{2s+1}$ by triangularizing the matrix (5) but this is expensive in terms of computing time. The crux of Francis' double QR procedure is that this transformation may be achieved very efficiently indeed.

Let

$$Q = Q_{2s}Q_{2s+1} \quad R = R_{2s+1}R_{2s}$$

so that

$$A_{2s}Q = QA_{2(s+1)}. \tag{6}$$

Now from the discussion in the previous chapter, we know that if we can determine an orthogonal matrix P which has the *same* first column as Q and an upper Hessenberg matrix H with positive subdiagonal elements such that

$$A_{2s}P = PH$$

then by uniqueness

$$Q = P \quad \text{and} \quad A_{2(s+1)} = H.$$

Likewise, if we can construct an orthogonal matrix P^T and an upper Hessenberg matrix H with positive subdiagonal elements such that P^T has the same first row as Q^T and such that

$$P^T A_{2s} = HP^T$$

then

$$Q = P \quad \text{and} \quad A_{2(s+1)} = H.$$

In order to construct such a matrix P, we require first of all to determine the first column of the matrix Q. Now by definition

$$Q^T(A_{2s}-k_s^{(1)}I)(A_{2s}-k_s^{(2)}I) = R$$

where the matrix $(A_{2s}-k_s^{(1)}I)(A_{2s}-k_s^{(2)}I)$ has the form (in the 6×6 case, for example)

$$\begin{bmatrix} x & x & x & x & x & x \\ x & x & x & x & x & x \\ x & x & x & x & x & x \\ 0 & x & x & x & x & x \\ 0 & 0 & x & x & x & x \\ 0 & 0 & 0 & x & x & x \end{bmatrix}$$

i.e. an upper Hessenberg matrix with an additional subdiagonal.

The matrix Q^T may be constructed by the usual Givens or Householder decomposition algorithms. It is easiest to see the structure of Q^T if we consider the Givens reduction. Our aim is to determine the first row of Q^T. Now Q^T is constructed as a product of plane rotations which annihilate the below diagonal elements of the matrix (5). Thus, in order to determine the first row of Q^T, we need only consider those rotations which affect the first row of Q^T. A consideration of Givens' procedure shows that such rotations must be of the form $R(1, p)$ for $p > 1$, and further that they are only used to annihilate the subdiagonal elements in the first column of the matrix $(A_{2s} - k_s^{(1)}I)(A_{2s} - k_s^{(2)}I)$. In fact, there are only two such elements and hence the first row of Q^T is fixed by considering only two plane rotations $R(1, 2)$ and $R(1, 3)$ which are chosen to annihilate the (2, 1), (3, 1) elements of the matrix (5) and to make the (1, 1) element positive. After determining the matrices $R(1, 2)$, $R(1, 3)$ we form the matrix G where

$$R(1, 3)\, R(1, 2)\, A_{2s} R^T(1, 2)\, R^T(1, 3) = G.$$

Since A_{2s} is of upper Hessenberg form it is clear that G will be of upper Hessenberg form in all but the first four rows.

We must now determine an orthogonal matrix S which transforms G into upper Hessenberg form and which in addition has its first column equal to $\mathbf{e}_1$. This latter condition is necessary to ensure that the first row of

$$S^T R(1, 3)\, R(1, 2)$$

is the same as that of Q^T. The matrix S is easily determined by the Givens procedure to satisfy the above conditions. Thus

$$S^T G S = H$$

where H is an upper Hessenberg matrix with positive subdiagonal elements. It follows that

$$\{S^T R(1, 3)\, R(1, 2)\}\, A_{2s} \{R^T(1, 2)\, R^T(1, 3)\, S\} = H$$

or that

$$P^T A_{2s} = H P^T.$$

But since P^T has the same first row as Q^T it follows that

$$P^T = Q^T \quad \text{and} \quad H = A_{2(s+1)}.$$

Thus the two steps of the QR process leading to $A_{2(s+1)}$ have been carried out using real arithmetic only.

This procedure is extremely efficient. For, the determination of the matrix G is carried out in a fixed number of operations independent of the order of the matrix A_{2s}. Moreover it may be verified that the transformation of G into H may be achieved in approximately $8n^2$ multiplications, which would be the same number required for two steps of a real QR algorithm on a Hessenberg matrix. Two steps of a complex QR algorithm require roughly twice as many multiplications. In practice, it is more efficient to use elementary Hermitian matrices to carry out the reduction of G to H, and even to determine G from A_{2s}.

Once again, the convergence of this algorithm depends on careful use of the deflating or decomposing strategies in the standard QR process. The eigenvectors may be obtained by the usual techniques.

A similar analysis to the last two sections could be carried out for the LR process. However, it leads to much less satisfactory algorithms than those based on the QR decomposition. The double QR algorithm is a very powerful technique for solving the real eigenproblem.

14

Generalized eigenvalue problems

14.1 Introduction

So far in this book, we have been exclusively concerned with the problem of finding values of λ and $\mathbf{x}$ such that

$$(A-\lambda I)\,\mathbf{x} = \mathbf{0} \tag{1}$$

has a non-trivial solution. This is the standard eigenproblem. Similar, though somewhat more general problems frequently arise in a number of engineering applications. For example (see also Sections 1.3 and 1.7) the equations of motion of a mechanical system under the application of a conservative system of forces can be written

$$A^{(0)}\frac{\mathrm{d}^2\mathbf{y}}{\mathrm{d}t^2} = -A^{(1)}\mathbf{y} \tag{2}$$

where $A^{(0)}$ and $A^{(1)}$ are symmetric and $A^{(0)}$ is positive definite, and $\mathbf{y}$ is a vector of displacements about a position of stable equilibrium.

Assuming a solution of the form

$$\mathbf{y} = \mathbf{x}\,\mathrm{e}^{i\mu t}$$

where $i = \sqrt{-1}$ and μ is a constant, then we have on substitution into equation (2) the equation

$$\mu^2 A^{(0)}\mathbf{x} = A^{(1)}\mathbf{x},$$

or

$$(A^{(0)}\lambda - A^{(1)})\,\mathbf{x} = \mathbf{0} \tag{3}$$

writing $\lambda = \mu^2$.

The problem is now a special case of the following: find values of λ and $\mathbf{x}$ such that

$$(A^{(0)}\lambda^r + A^{(1)}\lambda^{r-1} + \ldots + A^{(r)})\,\mathbf{x} = 0 \tag{4}$$

(where r is a positive integer) has a non-trivial solution. This is referred to as a *generalized eigenproblem*.

Provided that the matrix $A^{(0)}$ is non-singular, such a problem can be reduced to one of standard type in a straightforward manner. For if we define

$$\mathbf{x}_i = \lambda \mathbf{x}_{i-1}, \qquad i = 1, 2, \ldots, r-1$$

with

$$\mathbf{x}_0 = \mathbf{x},$$

and

$$B^{(i)} = -A^{(0)-1}A^{(i)}, \quad i = 1, 2, \ldots, r,$$

we have

$$\begin{bmatrix} 0 & I & 0 & \cdots & \cdots & 0 \\ 0 & 0 & I & \cdots & \cdots & 0 \\ \cdot & \cdot & & & & \cdot \\ \cdot & \cdot & & & & \cdot \\ \cdot & \cdot & & & & \cdot \\ 0 & 0 & 0 & \cdots & \cdots & I \\ B^{(r)} & B^{(r-1)} & B^{(r-2)} & \cdots & \cdots & B^{(1)} \end{bmatrix} \begin{bmatrix} \mathbf{x}_0 \\ \mathbf{x}_1 \\ \cdot \\ \cdot \\ \cdot \\ \mathbf{x}_{r-2} \\ \mathbf{x}_{r-1} \end{bmatrix} = \lambda \begin{bmatrix} \mathbf{x}_0 \\ \mathbf{x}_1 \\ \cdot \\ \cdot \\ \cdot \\ \mathbf{x}_{r-2} \\ \mathbf{x}_{r-1} \end{bmatrix} \tag{5}$$

Since the matrix on the left-hand side of this equation is sparse, iterative methods can profitably be used for the solution of this problem, and in particular inverse iteration is useful if good approximations are known to the eigenvalues. Because of the possible size of the system, and also because the matrix $A^{(0)}$ may not be well-conditioned with respect to inversion (i.e. small changes in its elements may lead to large changes in the elements of the inverse), it is often more efficient, however, to work with the original equation (4). For example provided that good approximations are known to the eigenvalues, iterative methods as described in Chapter 12 may be applied to determine the nr zeros of

$$\det\,(A^{(0)}\lambda^r + A^{(1)}\lambda^{r-1} + \ldots + A^{(r)}).$$

This approach is particularly useful for the problems considered in the next section.

It is sometimes possible, however, to apply inverse iteration to equation (5) in an extremely efficient manner. To illustrate, consider the case $r = 2$. Then the iterative procedure (without normalization) is given by

$$\begin{bmatrix} -\mu I & I \\ B^{(2)} & B^{(1)} - \mu I \end{bmatrix} \begin{bmatrix} \mathbf{x}_0^{(s+1)} \\ \mathbf{x}_1^{(s+1)} \end{bmatrix} = \begin{bmatrix} \mathbf{x}_0^{(s)} \\ \mathbf{x}_1^{(s)} \end{bmatrix}$$

where μ is an approximation to an eigenvalue. Thus

$$-\mu\mathbf{x}_0^{(s+1)}+\mathbf{x}_1^{(s+1)} = \mathbf{x}_0^{(s)} \tag{6}$$

and

$$B^{(2)}\mathbf{x}_0^{(s+1)}+(B^{(1)}-\mu I)\,\mathbf{x}_1^{(s+1)} = \mathbf{x}_1^{(s)}. \tag{7}$$

Eliminating $\mathbf{x}_0^{(s+1)}$ between these two equations, we have

$$(B^{(2)}+\mu B^{(1)}-\mu^2 I)\,\mathbf{x}_1^{(s+1)} = B^{(2)}\mathbf{x}_0^{(s)}+\mu\mathbf{x}_1^{(s)},$$

i.e.

$$(\mu^2 A^{(0)}+\mu A^{(1)}+A^{(2)})\,\mathbf{x}_1^{(s+1)} = A^{(2)}\mathbf{x}_0^{(s)}-\mu A^{(0)}\mathbf{x}_1^{(s)}. \tag{8}$$

Thus $x_1^{(s+1)}$ and $x_0^{(s+1)}$ can be computed from equations (8) and (6) respectively. Only one $n\times n$ system of linear equations need be solved at each stage.

We now turn to some non-standard eigenvalue problems whose matrices have special properties deriving from the physical situation from which they arise.

14.2 Parameterized matrices

An important special case of the generalized eigenvalue problem arises when the elements of the matrices $A^{(i)}$ are functions of an independent parameter α. Such problems arise, for example, in the modelling of aerodynamical systems, and involve tracing the history of certain of the eigenvalues as α is allowed to vary. If a solution is obtained for some initial value α_0, and then a small perturbation introduced into α_0 to give α_1, we can expect that the computed values give good approximations to the new set of eigenvalues. In this case, there are clearly advantages to be gained in working directly with

$$\det\,(A^{(0)}\lambda^r+A^{(1)}\lambda^{r-1}+\ldots+A^{(r)})$$

and using an iterative method for finding the nr zeros.

The efficient evaluation of this determinant is desirable, and this can be obtained from the triangular factors given by Gaussian elimination with pivoting.

14.3 The eigenvalue problem $A\mathbf{x} = \lambda B\mathbf{x}$

As we have seen, certain engineering problems [see also, for example, Gupta (1969, 1970)] give rise to eigenvalue problems of the form

$$A\mathbf{x} = \lambda B\mathbf{x}, \tag{9}$$

where A is symmetric, and B is symmetric and positive definite. In general, this problem is most efficiently solved by reduction to a standard symmetric eigenvalue problem in the following manner. Let

$$B = LL^T$$

be the Cholesky decomposition of B, where L is a real, non-singular lower triangular matrix. This decomposition is extremely stable. Then equation (9) is equivalent to

$$C\mathbf{y} = \lambda\mathbf{y} \tag{10}$$

where

$$C = L^{-1}A(L^{-1})^T$$

and

$$\mathbf{y} = L^T\mathbf{x}.$$

Frequently, both the matrices A and B have zero elements outside a band down the main diagonal. If the band width is sufficiently small, it is important that we take full advantage of this. Reduction of the original problem to one of the form (10) means that the band structure is destroyed, since C will in general be a full matrix, and we now describe a more efficient procedure for use under these circumstances.

We note firstly that since C is symmetric, the matrix $C-\lambda I$ possesses the Sturm sequence property described in Chapter 9. We now show that the matrix $A-\lambda B$ possesses the same property. If A_r, B_r and C_r denote respectively the leading principal minors of order r of A, B and C, we require to show that the signs of $\det(A_r-\lambda B_r)$ and $\det(C_r-\lambda I)$ are the same, for all r.

Let B be decomposed as above and let L_r be the leading principal sub-matrix of order r of L. Then we have

$$L_rL_r^T = B_r$$

$$L_r^{-1}A_r(L_r^{-1})^T = C_r.$$

Thus

$$\begin{aligned}\det(A_r-\lambda B_r) &= \det(L_r(L_r^{-1}A_r(L_r^{-1})^T-\lambda I)\,L_r^T)\\ &= \det(L_r(C_r-\lambda I)\,L_r^T)\\ &= \det(L_r)^2\det(C_r-\lambda I).\end{aligned}$$

Since L_r is non-singular, it follows that the signs of $\det(A_r-\lambda B_r)$ and $\det(C_r-\lambda I)$ are the same. Consequently, the number of eigenvalues of $A-\lambda B$ greater than λ is equal to the number of agreements in sign between consecutive members of the sequence

$$\det(A_r-\lambda B_r), \quad r = 0, 1, 2, \ldots, n,$$

with $\det(A_0-\lambda B_0) = 1$ by definition, and individual eigenvalues can be obtained by the method of bisection described in Section 9.3.

It remains to consider the economical evaluation of the relevant determinants, and this is conveniently done from the triangular factors of $A-\lambda B$. The Cholesky decomposition corresponding to equation (10) cannot be used, as in general $A-\lambda B$ is not positive definite, and the stable decomposition methods described in Section 3.2 require that the rows be permuted. However, a modification of this latter procedure can be used, in which the reduction process involved in triangular decomposition with pivoting is applied to only the first $r+1$ rows, for successive values of r from 1 to $n-1$.

Let $D = A-\lambda B$. Then the factorization involves $(n-1)$ major steps for an $n\times n$ matrix D, with only the first $(r+1)$ rows being involved in the rth step. For a full matrix, the rth major step involves the following operations, which are performed for each value of i from 1 to r:

(i) Interchange $d_{r+1,j}$ and d_{ij} for $j = i, i+1, \ldots, n$ if $|d_{r+1,i}|>|d_{ii}|$.
(ii) Replace $d_{r+1,j}$ by $d_{r+1,j}-d_{ij}d_{r+1,i}/d_{ii}$ for $j = i+1, i+2, \ldots, n$ (where d_{ij} denotes the element currently in position (i, j) of D).

The determinant of the $(r+1)$st principal minor is now given by

$$\det(D_{r+1}) = (-1)^N d_{11}d_{22}\ldots d_{r+1,r+1}, \tag{11}$$

where N is the total number of interchanges which have been performed to this point. In fact, since only the signs of the determinants are required, the multiplications in equation (11) need not actually be carried out.

If the matrix D has band structure, then the above operations need not be performed for all i from 1 to r, but only for those i for which $d_{r+1,i}$ is non-zero.

Example Evaluate the determinants of the principal minors by the above process of the tridiagonal matrix

$$D = \begin{bmatrix} 1 & 4 & & \\ 4 & 2 & 1 & \\ & 1 & 3 & 4 \\ & & 4 & 3 \end{bmatrix}$$

(*We assume rounding to four decimal places when required.*)

$\det(D_1) = 1$

$r = 1$

$$\xrightarrow{\textit{Step 1}} \begin{bmatrix} 4 & 2 & 1 & \\ 1 & 4 & & \\ & 1 & 3 & 4 \\ & & 4 & 3 \end{bmatrix}$$

$$\xrightarrow{\textit{Step 2}} \begin{bmatrix} 4 & 2 & 1 & \\ & 3{\cdot}5 & -0{\cdot}25 & \\ & 1 & 3 & 4 \\ & & 4 & 3 \end{bmatrix}$$

$\det(D_2) = (-1).\,4(3{\cdot}5) = -14{\cdot}0$

$r = 2$ (no interchanges required)

$$\xrightarrow{\textit{Step 2}} \begin{bmatrix} 4 & 2 & 1 & \\ & 3{\cdot}5 & -0{\cdot}25 & \\ & 0 & 3{\cdot}0714 & 4 \\ & & 4 & 3 \end{bmatrix}$$

$\det(D_3) = \det(D_2)\,(3{\cdot}0714) = -42{\cdot}9996$

$r = 3$

$$\xrightarrow{\textit{Step 1}} \begin{bmatrix} 4 & 2 & 1 & \\ & 3{\cdot}5 & -0{\cdot}25 & \\ & & 4 & 3 \\ & & 3{\cdot}0714 & 4 \end{bmatrix}$$

$$\xrightarrow{\textit{Step 2}} \begin{bmatrix} 4 & 2 & 1 & \\ & 3{\cdot}5 & -0{\cdot}25 & \\ & & 4 & 3 \\ & & & 1{\cdot}6965 \end{bmatrix}$$

$\det(D_4) = -\det(D_2)\,(4)\,(1{\cdot}6965) = 95{\cdot}0040$

The Sturm sequence property, together with the method of bisection, can be used to locate any (or all) of the eigenvalues of $A\mathbf{x} = \lambda B\mathbf{x}$. If the band width is fairly large, it is desirable to evaluate the Sturm sequence as small a number of times as possible, and Peters and Wilkinson (1969) describe an efficient implementation of this algorithm incorporating a process based on linear interpolation.

We conclude this section with a warning that the eigenvalue problem (9) can be ill-conditioned if B is ill-conditioned with respect to inversion. For example let

$$A = \begin{bmatrix} 1 & 2 \\ 2 & 0 \end{bmatrix}, \quad B = \begin{bmatrix} 1+\epsilon & 1 \\ 1 & 1 \end{bmatrix}$$

where $\epsilon > 0$. Then we have

$$\det(A - \lambda B) = \epsilon\lambda^2 + 3\lambda - 4,$$

and the eigenvalues are

$$\lambda = \frac{-3 \pm \sqrt{9+16\epsilon}}{2\epsilon}.$$

As $\epsilon \to 0$, the eigenvalues tend to 4/3 and ∞.

Exercises

1. If A and B are band matrices, show that if the problem (9) is reduced to that given by equation (10), the band structure is destroyed.

2. If the algorithm described above is applied to a matrix with band width $2b+1$, show that the work involved in computing the Sturm sequence for one value of λ is approximately $2nb^2$ multiplications.

14.4 The eigenvalue problem $AB\mathbf{x} = \lambda\mathbf{x}$

Another frequently occurring non-standard eigenvalue problem is that defined by the equation

$$AB\mathbf{x} = \lambda\mathbf{x} \tag{12}$$

where A and B are symmetric, and one or both are positive definite. Although not strictly speaking a member of the class of generalized eigenvalue problems defined by equation (4), it has similarities with the eigenvalue problem dealt with in the last section.

It is in general undesirable to form the product AB, as this is not necessarily symmetric, and the problem can readily be posed as a symmetric eigenvalue problem. Further, this can be done in fewer operations than are required to form the product of A and B, and in a very stable fashion. For, if B is positive definite, then we can write

$$B = LL^T \tag{13}$$

where L is real non-singular and lower triangular. Thus

$$ALL^T\mathbf{x} = \lambda\mathbf{x}$$

giving

$$D\mathbf{y} = \lambda\mathbf{y}$$

where

$$D = L^TAL$$

and

$$\mathbf{y} = L^T\mathbf{x}.$$

Now, the Cholesky decomposition (13), as already pointed out, is extremely stable, and requires only $\frac{1}{6}n^3$ multiplications. Also if full advantage is taken of symmetry, the products AL and $L^T(AL)$ can be computed in a total of $\frac{1}{2}n^3$ multiplications. Thus the matrix D is formed in a total of $\frac{2}{3}n^3$ multiplications. The verification of these operation counts is left as an exercise.

Exercise

3. Solve, using the methods of this chapter, the eigenvalue problems (*a*) $A\mathbf{x} = \lambda B\mathbf{x}$ and (*b*) $AB\mathbf{x} = \lambda\mathbf{x}$
where

$$A = \begin{bmatrix} 1 & 2 & -1 \\ 2 & 5 & 0 \\ -1 & 0 & 1 \end{bmatrix}$$

and

$$B = \begin{bmatrix} 1 & 2 & 1 \\ 2 & 5 & 4 \\ 1 & 4 & 6 \end{bmatrix}.$$

References

Becker, J., and R. M. Hayes (1967), *Information Storage and Retrieval*, Wiley.

Clint, M., and A. Jennings (1970), 'The evaluation of eigenvalues and eigenvectors of real symmetric matrices by simultaneous iteration,' *Computer Journal* **13**, 76–80.

Dekker, T. (1968), Algol 60 procedures in numerical algebra Part I, Mathematical Centrum Tracts No. 22. Amsterdam.

Dekker, T. (1970), Algol 60 procedures in numerical algebra Part II, Mathematical Centrum Tracts No. 23. Amsterdam.

Dernburg, T. F., and J. D. Dernburg (1969), *Macroeconomic Analysis*, Addison-Wesley.

Eberlein, P. J. (1962), A Jacobi-like method for the automatic computation of eigenvalues and eigenvectors of an arbitrary matrix, *Journal of the Society for Industrial and Applied Mathematics* **10**, 74–88.

Forsythe, G., and C. Moler (1967), *Computer Solution of Linear Algebraic Systems*, Prentice Hall.

Goldstine, H. H., and L. P. Horwitz (1959), 'A procedure for the diagonalization of normal matrices,' *Journal of the Association for Computing Machinery* **6**, 176–195.

Greenstadt, J. (1955), 'A method for finding roots of arbitrary matrices,' *Mathematical Tables and Aids to Computation* **9**, 47–52.

Gupta, K. K. (1969), 'Free vibrations of single branch structural systems,' *Journal of the Institute of Mathematics and its Applications* **5**, 351–362.

Gupta, K. K. (1970), 'Vibration of frames and other structures with banded stiffness matrix,' *International Journal for Numerical Methods in Engineering* **2**, 221–228.

Lotkin, M. (1956), 'Characteristic values of arbitrary matrices,' *Quarterly Journal of Applied Mathematics* **14**, 267–275.

Peters, G., and J. H. Wilkinson (1969), 'Eigenvalues of $A\mathbf{x} = \lambda B\mathbf{x}$ with banded symmetric A and B,' *Computer Journal* **12**, 398–404.

Rutishauser, H. (1969), 'Computational aspects of F. L. Bauer's simultaneous iteration method', *Numerische Mathematik* **13**, 4–13.

Traub, J. F. (1964), *Iterative Methods for the Solution of Equations*, Prentice-Hall.

Wilkinson, J. H. (1965), *The Algebraic Eigenvalue Problem*, Oxford University Press.

Wilkinson, J. H. (1968), 'Global convergence of tridiagonal QR algorithm with origin shifts, *Linear Algebra and its Applications* **1**, 409–420.

Wilkinson, J. H., and C. Reinsch (1971), *Handbook for Automatic Computation, Vol. II Linear Algebra*, Springer-Verlag.

Index